Mixed-Signal Systems

IEEE Press
445 Hoes Lane, P.O. Box 1331
Piscataway, NJ 08855-1331

Mixed-Signal Systems

A Guide to CMOS Circuit Design

Andrzej Handkiewicz

IEEE

IEEE Press

WILEY-INTERSCIENCE

A JOHN WILEY & SONS, INC., PUBLICATION

This text is printed on acid-free paper. ♾

Published simultaneously in Canada.

For ordering and customer service, call 1-800-CALL WILEY.

Library of Congress Cataloging-in-Publication Data is available.

ISBN 978-0-471-22853-0

In memory of my father

Contents

List of Figures

List of Tables

Preface

Faith and reason are like two wings
on which the human spirit rises
to the contemplation of truth; . . .

John Paul II, *Fides et Ratio*

Advances in VLSI technology have allowed designers to integrate digital and analog circuits on a single chip. This possibility is very useful in signal processing; signals are taken from the analog world and processed with digital systems. These signals require the use of analog pre- and postprocessing blocks such as converters, filters, sensors, drivers, buffers, and actuators. There exist many digital-oriented design systems, so-called silicon compilers whose common features are presented in the book. However, the lack of universal computer tools for the synthesis, simulation, and layout of the analog part on the chip is a design bottleneck of modern VLSI circuits. This book is devoted to design methods of both digital and analog integrated circuits. Some design methods and computer tools that are described in the book are not yet applied in any compiler. Hence, this work can be a source of information not only for graduate students and integrated circuit designers, but also for developers of silicon compilers.

There are many computer systems such as Cadence, Saber, Compass, Mentor Graphics, and Synopsis that can be helpful in the design of integrated circuits. There are books and manuals describing each system. This book is more general in character. It does not describe any particular system. Part II of the book presents the basic methods, procedures, and tools used in silicon compilers. The Part II is preceded by Part I, in which basics that are necessary for a user of a particular compiler are described. Part I presents basic information about CMOS technology. The techniques most frequently used and fully implemented in this technology are also included in Part I.

Each author faces the dilemma of which subjects can be omitted and which must be included in his book. Such choice was especially difficult in this book, which is on the border of three main areas: signal processing, systems on a chip, and computer tools. If too many problems and approaches were considered, the book would be unclear and difficult to absorb for the reader. Hence, the reader will not find separated, detailed descriptions of CMOS technology, CMOS circuits techniques, and computer tools for synthesis and design of these circuits in this book. The book is focused on the problems in these areas, which are strictly connected each other.

The book is composed of eight chapters and divided into two parts. The chapters that describe analog and discrete signals, basics of CMOS technology, and basic cells that can be obtained in digital, OTA-C, switched-capacitor, and switched-current techniques are included into Part I. Part II is devoted to methods and tools that can be used to implement mixed-signal systems on a chip.

In the first chapter, we review topics concerning analog and digital signals, which are continuous- and discrete-time functions. Notation and description of mixed-signals is briefly presented. The topics are: Fourier, $\mathcal{Z}$, and Laplace transforms, the sampling theorem, and the aliasing problem. Exponential, Euler, and bilinear mappings from s to z domains are also presented. Transfer functions describing two-dimensional systems in both domains are considered. The discrete cosine transform, which is very important in image compression and is used in the second part of the book, is also described in this chapter.

The second chapter describes the properties of MOS transistors used to realize all basic cells of integrated circuits. Operation, performance, and design of these cells are also presented in this chapter.

The third chapter contains brief description of the basic CMOS processes: wafer preparation, oxidation, deposition, litography and etching, epitaxy, diffusion, and ion implantation. The designer of integrated circuits ought to be familiar with these processes in order to understand the project steps and the design rules and to be able to modify the operations that violate these rules.

The fourth chapter describes digital techniques that are very frequently used in mixed-signal integrated circuit design. In particular, we describe static and dynamic logic gates, finite-state machines, and memories.

The fifth chapter presents the basic elements of passive and active circuits. We describe various implementations of integrators, i.e., switched-current, switched-capacitor and OTA-C implementations. These integrators are basic cells of mixed-signal processing systems. We also consider low-voltage and low-power operation of these elements.

The sixth chapter describes the low-sensitivity strategy in the design of one-dimensional filters implemented in switched-current, switched-capacitor, and OTA-C techniques. The gyrator introduced in the previous chapter plays a special role in this strategy. The chapter shows how to realize low-sensitivity strategy in digital filter design. Analog-to-digital converters based on delta-sigma modulation, which can be implemented in switched-current and switched-capacitor techniques, are also presented in this chapter.

In the next chapter we describe the architecture of a system on a chip, applicable for image processing. We assume that image sensing arrays are built on the chip in the standard CMOS process. Switched-current circuits are used for image preprocessing in the analog part of the chip.

Software systems that can automatically generate integrated circuits are called silicon compilers. The most popular systems are oriented on digital integrated circuit designs that are advantageous in the design, based on the hierarchy of the cells and the libraries of standard cells. Very effective computer systems for design of digital circuits exist, and tools that can be used in mixed-signal integrated circuit design are being intensively developed. The last chapter briefly presents the basic ideas that are necessary for the use and development of such systems.

Each chapter describes topics that are the background to further issues. A reader who is familiar with some of these topics can omit them. In order to help navigate the book, the most important connections between the chapters can be summarized out as:

- Euler and bilinear mappings in Chapter 1 are basic to the integrators described in Chapter 5 and the SI, SC, and digital filters described in Chapter 6.
- 2-D SI filters described in Chapter 7 are based on the novel 2-D multiport network description given in Chapter 1.
- Chapter 2 describes basic CMOS cells, which are used in Chapters 4 and 5.
- Chapter 3 is useful for understanding the design steps described in Chapters 7 and 8.
- Each digital part of the system on a chip presented in Chapter 7 is based on digital cells described in Chapter 4. The clock that is also described in this chapter is often used to control SC and SI circuits presented in Chapters 5, 6, and 7.
- The SC and SI cells described in Chapter 5 are used for the signal processing described in Chapters 6 and 7.
- Chapters 6 and 7 are focused on filters and other macrocells of mixed-signal processing systems.
- Chapter 8 explains the role of computer tools that are helpful in the design of systems presented in Chapters 6 and 7.

Most chapters contain novel problems that the reader could not find in existing books or overview articles. The novel topics include description of transfer functions of 2-D multiport networks (Chapter 1), novel SI memory and SI integrator cells (Chapter 5), implementation of the Darlington model of synthesis based on gyrator–capacitor multiport networks (Chapter 6), synthesis method of 2-D SI and SC filters (Chapter 7), and computer tools based on topological methods (Chapter 8).

The chapters contain problems that can be used as exercises for graduate students. The book presents an approach to one- and two-dimensional filter design based on the gyrator–capacitor prototype network. This approach is illustrated by image processing systems for implementation on a single chip and shows the reader the way from synthesis methods to silicon assembly.

I am grateful to my colleagues and students, especially to my Ph.D. students Marcin Łukowiak and Marek Kropidłowski for discussion and help in preparing the book and to Mrs. Krystyna Ciesielska, who spent much time helping me in English. I wish to thank my wife Hanna my daughter Magda, and my sons Maciej and Paweł for their understanding and support. I would like to express my gratitude to Maciej for his work on figures.

I sincerely hope that this book will be helpful to readers. I will be very grateful for comments and opinions.

ANDRZEJ HANDKIEWICZ

Andrzej.Handkiewicz@put.poznan.pl

Part I

Mixed Signals, Technology and Techniques

1

Continuous and Discrete Signals

In this chapter we shall review several concepts concerning analog and digital signals, namely the Fourier, $\mathcal{Z}$, and Laplace transforms, the sampling theorem, and the aliasing problem. These topics are presented in order to establish notation that we will use in mixed signal circuits. We will also present exponential, Euler, and bilinear mappings from the s domain to the z domain, as well as transfer functions describing two-dimensional systems in both domains. Finally, we will describe the discrete cosine transform, which is very important in image compression and will be used in the second part of the book.

1.1 FOURIER, $\mathcal{Z}$, AND LAPLACE TRANSFORMS

A discrete-time signal is defined as a sequence $\{x(k)\}$ resulting from sampling a continuous-time signal $x(t)$. The symbol $x(k)$ denotes the element of the sequence that is equal to the value of the function $x(t)$ for $t = kT$, where T is the sampling interval. The relation

$$x(k) = \int_{-\infty}^{\infty} \hat{x}_k(t)dt \tag{1.1}$$

describes the sampling operation, where

$$\hat{x}_k(t) = x(t)\delta(t - kT) \tag{1.2}$$

$\delta(t)$ is the delta function or distribution function. The function obtained as a sum of (1.2) for all indices k

$$\hat{x}(t) = \sum_k \hat{x}_k(t) = x(t)\sum_k \delta(t - kT) = \sum_k x(k)\delta(t - kT) \tag{1.3}$$

is called a continuous-time PAM (pulse amplitude modulation) representation of a discrete-time signal.

The periodic function $x(t)$ with a period T_p can be expanded in a Fourier series in the following complex form

$$x(t) = \sum_{n=-\infty}^{\infty} c_n e^{j(2\pi/T_p)nt} \tag{1.4}$$

where

$$c_n = \frac{1}{T_p}\int_{-T_p/2}^{T_p/2} x(t)e^{-j(2\pi/T_p)nt}dt \tag{1.5}$$

and $j = \sqrt{-1}$. The coefficients c_n fulfill the relation

$$c_n = c^*_{-n} = \frac{a_n - jb_n}{2} \tag{1.6}$$

where a_n, b_n, $n = 1, 2, 3, \ldots$, denote the coefficients of the Fourier series in a trigonometric form.

An extension of the formulae (1.4) and (1.5) for $T_p \to \infty$ gives the Fourier transform pair of an arbitrary continuous-time signal $x(t)$ in the form

$$X(j\omega) = \int_{-\infty}^{\infty} x(t)e^{-j\omega t}dt, \qquad x(t) = \frac{1}{2}\pi \int_{-\infty}^{\infty} X(j\omega)e^{j\omega t}d\omega \tag{1.7}$$

For the signal $\{x(k)\}$ the Fourier transform is called a discrete-time one (DTFT) and takes the form

$$X(e^{j\omega T}) = \sum_k x(k)e^{-j\omega kT}, \qquad x(k) = \frac{T}{2\pi}\int_{-\pi/T}^{\pi/T} X(e^{j\omega T})e^{j\omega kT}d\omega \tag{1.8}$$

If, however, in the above equations only N samples $x(k)$ are taken for $k = 0, 1, \ldots, N - 1$ and only N samples of $X(e^{j\omega T})$ are calculated for $\omega = n\omega_0$, $n = 0, 1, \ldots, N - 1$, where $\omega_0 = (2\pi/T)/N = \omega_s/N$, the discrete Fourier transform (DFT) is defined as

$$X_N(n) = X(e^{jn\omega_0 T}) = \sum_{k=0}^{N-1} x(k)e^{-j2\pi nk/N} \tag{1.9}$$

We see that the Fourier transform (1.7) gives relations between functions of real variables t and k and the frequency variable ω, which is also a real variable. In the case of the Laplace and Z transforms, after transformation of $x(t)$ and $x(k)$ we obtain functions

$$X(s) = \int_0^{\infty} x(t)e^{-st}dt \tag{1.10}$$

and

$$X(z) = \sum_{k=0}^{\infty} x(k)z^{-k} \tag{1.11}$$

of complex variables s and z, respectively. We assume that the functions $x(t)$ and $x(k)$ in (1.10), (1.11), are equal to zero for negative arguments t and $k[x(t) = 0$ for $t < 0$ and $x(k) = 0$ for $k < 0]$, i.e., that they are causal functions. For causal functions, the Laplace transform is equivalent to the Fourier transform for $s = j\omega$ and the Z transform to DTFT for $z = e^{j\omega T}$. It means that the variable ω is represented by the imaginary axis on the s plane and by the unit circle on the z plane. The Laplace transform (1.10) of the PAM representation of $x(k)$ described by (1.3) is as follows:

$$\hat{X}(s) - \sum_{k=0}^{\infty} x(k)e^{-skT} \tag{1.12}$$

For

$$z = e^{sT} \tag{1.13}$$

it gives the equivalence between the Laplace and Z transforms and the relation

$$\hat{X}(j\omega) = X(e^{j\omega T}) \tag{1.14}$$

Equation (1.13) shows that the imaginary axis on the s plane is transformed into the unit circle with the center in the origin of the coordinate system in the z domain. On the basis of this equation, we can add that the left-hand side of the s plane is transformed into the interior of this circle, whereas the right-hand side is transformed into the exterior. For the z^{-1} plane

$$z^{-1} = e^{-sT} \tag{1.15}$$

which is also often considered. The relations between the left- and right-hand sides on the s plane and the interior and exterior of the unit circle in the z^{-1} domain are reversed.

The Z, Fourier, and Laplace transforms of functions corresponding to basic signals are shown in Table 1.1 Function $u(t)$ denotes the unit step.

Table 1.1 Transforms of basic signals

$x(t)$	$Z\{u(nT)x(nT)\}$	$\mathcal{F}\{x(t)\}$	$\mathcal{L}\{u(t)x(t)\}$
$\delta(t)$	1	1	1
$u(t)$	$\frac{z}{z-1}$	$\pi\delta(\omega)+\frac{1}{j\omega}$	$\frac{1}{s}$
$sgn(t)$	$\frac{z}{z-1}$	$\frac{2}{j\omega}$	$\frac{1}{s}$
$\Pi_{2k}(t)$	$\frac{z(1-z^{-k})}{z-1}$	$\frac{2\sin k\omega}{\omega}$	$\frac{1-e^{-ks}}{s}$
$e^{j\omega_o t}$	$\frac{z^2-ze^{-j\omega_o T}}{z^2-2z\cos\omega_o T+1}$	$2\pi\delta(\omega-\omega_o)$	$\frac{1}{s-j\omega_o}$
$u(t)e^{-\alpha t}$	$\frac{z}{z-e^{-\alpha T}}$	$\frac{1}{\alpha+j\omega}$	$\frac{1}{s+\alpha}$
$e^{-\alpha\|t\|}$	$\frac{z}{z-e^{-\alpha T}}$	$\frac{2\alpha}{\alpha^2+\omega^2}$	$\frac{1}{s+\alpha}$
$\cos\omega_o t$	$\frac{z^2-z\cos\omega_o T}{z^2-2z\cos\omega_o T+1}$	$\pi\delta(\omega-\omega_o)+\pi\delta(\omega+\omega_o)$	$\frac{s}{s^2+\omega_o^2}$
$u(t)e^{-\alpha t}\cos\beta t$	$\frac{z^2-ze^{-\alpha T}\cos\beta T}{z^2-2ze^{-\alpha T}\cos\beta T+e^{-2\alpha T}}$	$\frac{\alpha+j\omega}{(\alpha+j\omega)^2+\beta^2}$	
$\sin\omega_o t$	$\frac{z\sin\omega_o T}{z^2-2z\cos\omega_o T+1}$	$j\pi\delta(\omega+\omega_o)-j\pi\delta(\omega-\omega_o)$	$\frac{\omega_o}{s^2+\omega_o^2}$
$u(t)e^{-\alpha t}\sin\beta t$	$\frac{ze^{-\alpha T}\sin\beta T}{z^2-2ze^{-\alpha T}\cos\beta T+e^{-\alpha T}}$	$\frac{\beta}{(\alpha+j\omega)^2+\beta^2}$	$\frac{\beta}{(s+\alpha)^2+\beta^2}$

1.2 ALIASING PHENOMENON AND NYQUIST SAMPLING THEOREM

A linear, time-invariant (LTI) system excited by the signal $x(t)$ responds with a continuous-time signal $y(t)$. For the delta excitation $[x(t) = \delta(t)]$ the response is denoted by $h(t)$ and called the pulse response. Any response $y(t)$ of the LTI system can be expressed in the time domain as a convolution:

$$y(t) = x(t) * h(t) = \int_{-\infty}^{\infty} x(\tau)h(t-\tau)d\tau \tag{1.16}$$

or as a multiplication

$$Y(s) = H(s)X(s), \qquad Y(j\omega) = H(j\omega)X(j\omega) \tag{1.17}$$

in the Laplace and Fourier domains. Similarly, for a discrete-time system we have

$$y_k = x(k) * h_k = \sum_{m=-\infty}^{\infty} x_m h_{k-m} \tag{1.18}$$

in the discrete-time domain, or

$$Y(z) = H(z)X(z), \qquad Y(e^{j\omega T}) = H(e^{j\omega T})X(e^{j\omega T}) \tag{1.19}$$

in the $\mathcal{Z}$ and Fourier domains.

Multiplication in the function $\hat{x}(t)$ presented in the second form in (1.3)

$$\hat{x}(t) = [x(t)] \cdot \left[\sum_k \delta(t - kT)\right] \tag{1.20}$$

for LTI systems corresponds to convolution in the frequency domain

$$\begin{aligned}\hat{X}(j\omega) &= X(e^{j\omega T}) \\ &= \frac{1}{2\pi}[X(j\omega)] * \left[\omega_s \sum_{m=-\infty}^{\infty} \delta(\omega - m\omega_s)\right] \\ &= \frac{1}{T}\int_{-\infty}^{\infty} X(j\Omega) \sum_{m=-\infty}^{\infty} \delta(\omega - \Omega - m\omega_s)d\Omega \\ &= \frac{1}{T}\sum_{m=-\infty}^{\infty} X[j(\omega - m\omega_s)]\end{aligned} \tag{1.21}$$

It means that the Fourier transform $X(e^{j\omega T})$ of a discrete signal can be obtained as a sum of shifted Fourier transforms $X(j\omega)$ of a continuous-time signal [13]. Each component in this sum is shifted by the integer multiple m of the sampling frequency $\omega_s = 2\pi/T$. It means that the spectrum of the discrete signal can contain high-frequency components of $x(t)$ transposed to low-frequency components. This phenomenon is called aliasing. In order to eliminate aliasing, the signal $x(t)$ is fed to an ideal low-pass filter, called antialiasing filter, with the cutoff frequency $\omega_c \leq \omega_s/2$. In this case, there will be no overlap of frequency components of the signal sampled at the output of this filter. The continuous-time signal can be reconstructed again at the output of the next low-pass filter, called the smoothing filter, excited by a discrete signal. Hence, the signal $x(t)$ which has the Fourier transform $X(j\omega)$ and is sampled at frequency $2\pi/T$, can be reconstructed from its samples if $X(j\omega) = 0$ for all $|\omega| > \pi/T$, (Nyquist sampling theorem). The frequency $\omega_N = \pi/T$ is called the Nyquist frequency. The system for mixed signal processing, containing antialiasing and smoothing filters and presented in Figure 1.1, can be realized as a CMOS circuit on a single chip.

Using the ideal lowpass filter, which has the pulse response

$$h(t) = \frac{\sin(\pi t/T)}{\pi t/T} \tag{1.22}$$

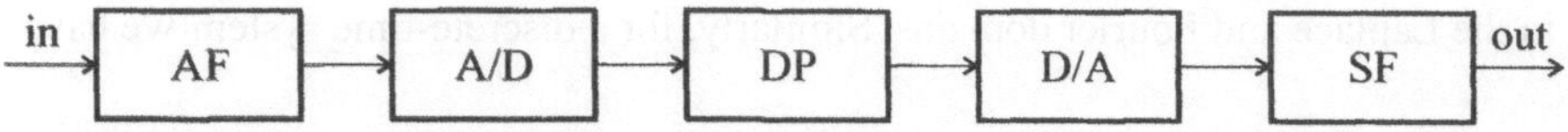

Figure 1.1 Example of a system for mixed signal processing composed of antialiasing (AF) and smoothing (SF) filters, A/D and D/A converters, and a digital core (DP).

we can obtain the analog signal $x(t)$ at the output of this filter excited by the PUM representation $\hat{x}(t)$ as the convolution $\hat{x}(t) * h(t)$. The reconstruction formula is as follows

$$x(t) = \hat{x}(t) * \left[\frac{\sin(\pi t/T)}{\pi t/T} \right] = \sum_{m=-\infty}^{\infty} x_m \frac{\sin[\pi(t - mT)/T]}{\pi(t - mY)/T} \tag{1.23}$$

1.3 EULER AND BILINEAR TRANSFORMATIONS

LTI systems are described by transfer functions that are rational functions in z and s domains. Discrete-time systems are often designed on the basis of continuous-time systems with the use of the transfer function $H(s)$. However, it is not possible to derive the rational transfer function $H(z)$ from $H(s)$ using the transformation (1.13). Hence, different approximations of relation (1.13) are used. The simplest ones result from the series expansion of exponential functions in the form

$$z = e^{sT} = 1 + \frac{sT}{1!} + \frac{(sT)^2}{2!} + \frac{(sT)^3}{3!} + \ldots \tag{1.24}$$

or

$$z^{-1} = e^{-sT} = 1 + \frac{-sT}{1!} + \frac{(-sT)^2}{2!} + \frac{(-sT)^3}{3!} + \ldots \tag{1.25}$$

and are called the forward and backward Euler transformations:

$$sT = z - 1 \tag{1.26}$$

and

$$sT = 1 - z^{-1} \tag{1.27}$$

respectively.

Another transformation, not so simple as the Euler transformations, but with very interesting properties, is the bilinear transformation

$$\frac{sT}{2} = \frac{z - 1}{z + 1} = \frac{1 - z^{-1}}{1 + z^{-1}} \tag{1.28}$$

or

$$z^{-1} = \frac{1 - sT/2}{1 + sT/2} \tag{1.29}$$

which can be obtained from the series representation of the ln function

$$sT = \ln(z) = 2\left[\frac{z-1}{z+1} + \frac{(z-1)^3}{3(z+1)^3} + \frac{(z-1)^5}{5(z+1)^5} + \ldots\right] \tag{1.30}$$

We see from relations (1.26) and (1.27) that the imaginary axis $s = j\omega$ in the s domain corresponds to the line tangent to the unit circle at the point (0, 1) on the z and z^{-1} planes, respectively. The left-hand side of the s plane corresponds to half-plane on the left-hand side of this tangent in the z domain and on the right-hand side in the z^{-1} domain. Let us note that the exact transformation (1.13) transforms the left-hand side of the s plane into the interior or exterior of the unit circle in the z and z^{-1} domains, respectively. For $s = j\omega$ the bilinear relation (1.29) yields $|z| = 1$ and, like the exact transformation (1.13), transforms the imaginary axis in the analog domain into the unit circle in the discrete domain. These relations between analog s and discrete z domains are shown in Figure 1.2.

The Euler and bilinear transformations impose scaling of frequencies ω_a and ω_d in analog and discrete domains. In the case of bilinear transformation, introducing into (1.28) the frequencies $s = j\omega_a$ and $z = e^{j\omega_d T}$, we obtain

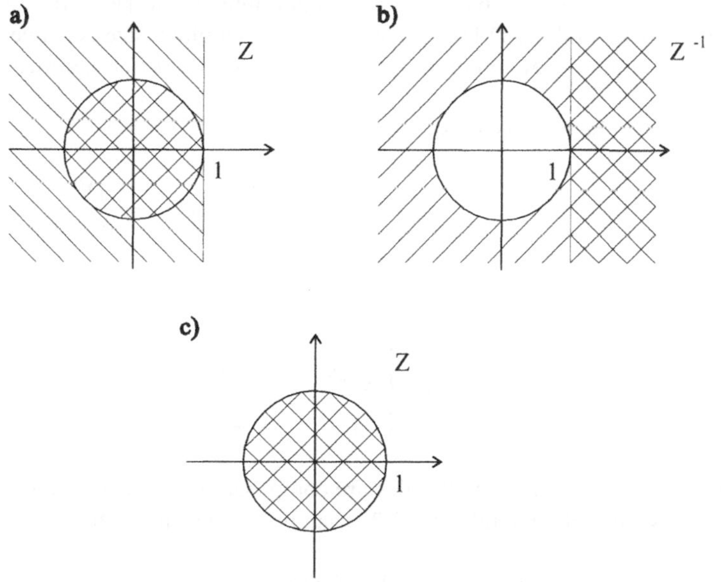

Figure 1.2 Transformations between analog and discrete domains for forward Euler (a), backward Euler (b), and bilinear (c) transformations.

$$\frac{\omega_a T}{2} = \tan\frac{\omega_d T}{2} \tag{1.31}$$

Let us note that this relation compresses the whole frequency axis in the analog domain into the frequency range limited by the Nyquist frequency $\omega_N = \pi/T$. This property makes discrete filters obtained on the basis of prototype analog filters more selective. On the other hand, the design process of discrete filters requires the analog filter to change its frequencies according to the relation (1.31), in order to obtain the desired frequencies in the counterpart discrete filter. This stage of the design process is called prewarping.

1.4 TWO-DIMENSIONAL DISCRETE COSINE TRANSFORM}

The Fourier transform presented in the previous sections can also be used for two-dimensional (2-D) processing. However, the optimum transform for image compression is the Karhunen–Loeve transformation (KLT) [31], because it packs the greatest amount of energy in the smallest number of elements in the frequency domain of a 2-D signal and minimizes the total entropy of the signal sequence. Unfortunately, the basis functions of KLTs are image-dependent, which is the most important implementation-related deficiency. It is observed that the two-dimensional (2-D) discrete cosine transform (DCT) has the output close to the output produced by the KLT [3], and uses image-independent basis functions. Hence, DCT-based image coding is applied in all video compression standards. In these standards, the image is divided into 8 × 8 blocks in the spatial domain and DCT transforms them into 8 × 8 blocks in the 2-D frequency domain. Such block size is convenient with respect to computational complexity. Larger sizes do not offer significantly better compression.

2-D DCT can be expressed as

$$y_{kl} = \frac{c(k)c(l)}{4}\sum_{i=0}^{7}\sum_{j=0}^{7} x_{ij}\cos\frac{(2i+1)k\pi}{16}\cos\frac{(2j+1)l\pi}{16} \tag{1.32}$$

where $k, l = 0, 1, \ldots, 7$ and

$$c(k) = \begin{cases} \frac{1}{\sqrt{2}}, & k = 0 \\ 1, & k \neq 0 \end{cases} \tag{1.33}$$

Assuming that the matrices Y and X are composed of elements y_{ij} and x_{ij}, $i, j = 0, 1, \ldots, 7$, respectively, the relation (1.32) can be also written in the matrix form as

$$Y = CXC^t \tag{1.34}$$

where the matrix of coefficients C is as follows:

$$C=\frac{1}{2}\begin{bmatrix} d & d & d & d & d & d & d & d \\ a & c & e & g & -g & -e & -c & -a \\ b & f & -f & -b & -b & -f & f & b \\ c & -g & -a & -e & e & a & g & -c \\ d & -d & -d & d & d & -d & -d & d \\ e & -a & g & c & -c & -g & a & -e \\ f & -b & b & -f & -f & b & -b & f \\ g & -e & c & -a & a & -c & e & -g \end{bmatrix} \tag{1.35}$$

$a = \cos(\pi/16)$, $b = \cos(2\pi/16)$, $c = \cos(3\pi/16)$, $d = \cos(4\pi/16)$, $e = \cos(5\pi/16)$, $f = \cos(6\pi/16)$, $g = \cos(7\pi/16)$.

The main property of 2-D DCT, with respect to implementation, is separability. On the basis of the matrix equation (1.34), written in the form

$$Y = Z'C', \qquad Z = X'C' \tag{1.36}$$

we can realize 2-D DCT with two 1-D ones. The matrix X denotes one input 8×8 block, and its transposition X' in the relation $Z = X'C'$ means that it is read out column by column. The matrix Z, containing intermediate results, is obtained with the use of 1-D DCT, and is saved in a memory array. Transposition of this matrix in the first equation in (1.36) means that the elements of Z obtained successively for the current block are memorized in row cells and for the previous block are read out from column cells of the memory array. The intermediate results are processed in the same way as the input matrix X, giving the output signal matrix Y. The implementation of a 2-D DCT processor will be presented in the second part of this book.

The matrix relations (1.36) can be expressed as

$$y_n = c(n)\sqrt{\frac{2}{N}}\sum_{k=0}^{N-1} x(k)\cos\frac{(2k+1)n\pi}{2N} \tag{1.37}$$

describing a 1-D DCT in an explicit form, where $n = 0, 1, \ldots, N-1$. Equation (1.37) can be used to show the relationship between DCT and DFT given by (1.9), [47]. On the basis of $x(k)$, a $2N$-point sequence ξ_k can be obtained as

$$\xi_k = \begin{cases} x(k), & 0 \le k \le N-1 \\ x_{2N-k-1}, & N \le k \le 2N-1 \end{cases} \tag{1.38}$$

Let us note that the second half of ξ_k for $k = N, \ldots, 2N-1$ is a mirror image of the first half of ξ_k for $k = 0, \ldots, N-1$. The $2N$-point DFT of ξ_k is, from the definition (1.9), given by

$$\begin{aligned} X_{2N}(n) &= \sum_{k=0}^{2N-1} \xi_k e^{-j2\pi nk/(2N)} \\ &= \sum_{k=0}^{N-1} x(k)e^{-j2\pi nk/(2N)} + \sum_{k=N}^{2N-1} x_{2N-k-1}e^{-j2\pi nk/(2N)} \end{aligned} \tag{1.39}$$

for $n = 0, \ldots, 2N - 1$. The first summation on the right-hand side of the above equation can be written as

$$\sum_{k=0}^{N-1} x(k)e^{-j2\pi nk/(2N)} = e^{jn\pi/(2N)} \sum_{k=0}^{N-1} x(k)e^{-j(2k+1)n\pi/(2N)} \tag{1.40}$$

whereas the second one can be written as

$$\begin{aligned}\sum_{k=N}^{2N-1} x_{2N-k-1}e^{-j2\pi nk/(2N)} &= \sum_{k=N-1}^{0} x(k)e^{-j2\pi n(2N-k-1)/(2N)} \\ &= e^{jn\pi(1-4N)/(2N)} \sum_{k=N-1}^{0} x(k)e^{j(2k+1)n\pi/(2N)} \\ &= e^{jn\pi/(2N)} \sum_{k=0}^{N-1} x(k)e^{j(2k+1)n\pi/(2N)}\end{aligned} \tag{1.41}$$

Introducing the results from (1.40) and (1.41) into (1.39), we obtain

$$X_{2N}(n) = 2e^{jn\pi/(2N)} \sum_{k=0}^{N-1} x(k) \cos \frac{(2k+1)n\pi}{2N} \tag{1.42}$$

for $n = 0, \ldots, 2N - 1$. Hence, the DCT transform y_n in (1.37) can be obtained from the $2N$-point DFT using the equation

$$y_n = \frac{c(n)}{\sqrt{2N}} e^{-jn\pi/(2N)} X_{2N}(n) \tag{1.43}$$

for $n = 0, \ldots, N - 1$.

1.5 TRANSFER FUNCTIONS OF A 2-D MULTIPORT NETWORK

Transfer functions H of LTI systems are often described in analog (s) or discrete (z) complex domains, as can be seen in relations (1.17) and (1.19). In this section, we will consider relations between transfer functions of a system described in different complex domains. The formulae that will be presented refer to two-dimensional systems. The corresponding relationships for one-dimensional systems can be easily obtained as a special case of formulae introduced for 2-D systems.

Transfer functions H^{mn}, $m = 1, \ldots, M$, $n = 1, \ldots, N$, of a 2-D LTI network are rational functions of two complex variables. H^{mn} is an element of the matrix H that describes a linear 2-D multiport network shown in Figure 1.3. N denotes the number of inputs, whereas M denotes the number of outputs. The elements of the input and output vector signals x and y are also functions of the complex variables. Each variable belongs to the s or z domain. Hence, there are four equivalent representa-

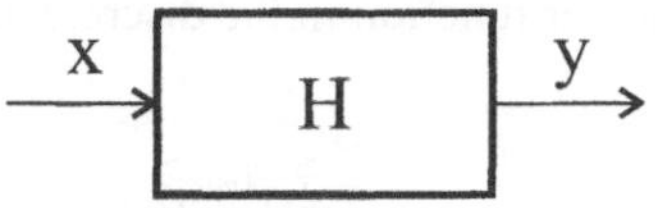

Figure 1.3 Symbol of a linear 2-D multiport network.

tions of the transfer functions H^{mn}, $m = 1, \ldots, M$, $n = 1, \ldots, N$: discrete, ($H^{mn} \leftarrow Z_1A^{mn}Z_2'/Z_1BZ_2'$), analog ($H^{mn} \leftarrow S_1P^{mn}S_2'/S_1QS_2'$), and hybrid ($H^{mn} \leftarrow S_1P_h^{mn}Z_2'/S_1Q_hZ_2'$) and ($H^{mn} \leftarrow Z_1A_h^{mn}S_2'/Z_1B_hS_2'$), as is shown in Figure 1.4, [21]. Let us note that each representation can have different numerators. However, they all have a common denominator.

The polynomials in the numerators and the denominator are written in the matrix form. The elements of matrices A^{mn}, B, P^{mn}, and Q are equal to coefficients of respective polynomials, and Z_i, S_i are vectors composed of complex variables z_i, s_i:

$$Z_i = [z_i^{-k} \cdots z_i^{-1}\ 1] \qquad S_i = [s_i^k \cdots s_i\ 1] \qquad i = 1, 2 \tag{1.44}$$

where the elements of the vectors Z_i, and S_i, $i = 1, 2$, are ordered in descending powers of variables z_i^{-1}, s_i, respectively. The sign $'$ denotes a transposed matrix or vec-

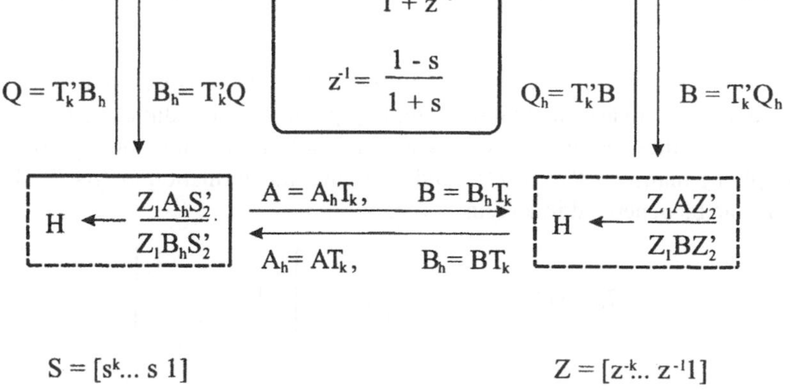

Figure 1.4 Representations of 2-D network transfer functions.

tor. Let us note that the transfer function in the discrete domain is usually written in the form

$$H^{mn} \leftarrow \frac{\tilde{Z}_1 \hat{A}^{mn} \tilde{Z}_2'}{\tilde{Z}_1 \hat{B} \tilde{Z}_2'} \tag{1.45}$$

where the elements of the vector $\tilde{Z}_i$ are ordered in ascending powers of the variable z_i^{-1}

$$\tilde{Z}_i = [1 z_i^{-1} \cdots z_i^{-k}] \tag{1.46}$$

and where $\hat{A}^{mn}$, $\hat{B}$ denotes the matrices A^{mn}, B transposed with respect to both diagonals. The description given by (1.45) is called the standard form of a transfer function.

The vectors S_1, S_2, Z_1, and Z_2 are used for describing polynomials in the numerators and denominators of the transfer functions. It does not mean that polynomials are of the same order with respect to the given variable because some rows or columns in the matrices A^{mn}, B, P^{mn}, and Q may be composed of zero elements. For example, the denominator of the transfer function of a nonrecursive filter is equal to 1 in the digital domain. One can describe this filter by the matrix B in the form

$$B = \begin{bmatrix} 0 & 0 & \cdots & 0 \\ 0 & 0 & \cdots & 0 \\ \vdots & \vdots & & \vdots \\ 0 & 0 & \cdots & 1 \end{bmatrix} \tag{1.47}$$

We assume that the discrete and analog variables are bilinearly transformed

$$s_i = \frac{1 - z_i^{-1}}{1 + z_i^{-1}}$$

$$z_i^{-1} = \frac{1 - s_i}{1 + s_i}, \qquad i = 1, 2 \tag{1.48}$$

The above relationships are obtained from (1.28) and (1.29) where, for the sake of simplicity, we will assume that the sampling periods in both dimensions $i = 1, 2$ are $T = 2$. Under these assumptions, we can obtain all transfer function representations, multiplying matrices A^{mn}, B, P^{mn}, and Q by the transformation matrix T_k. The matrix T_k can be generated in a recurrent manner:

$$T_0 = [1],\ T_1 = \begin{bmatrix} -1 & 1 \\ 1 & 1 \end{bmatrix},\ T_2 = \begin{bmatrix} 1 & -2 & 1 \\ -1 & 0 & 1 \\ 1 & 2 & 1 \end{bmatrix}$$

$$T_3 = \begin{bmatrix} -1 & 3 & -3 & 1 \\ 1 & -1 & -1 & 1 \\ -1 & -1 & 1 & 1 \\ 1 & 3 & 3 & 1 \end{bmatrix}, \ldots, T_k \tag{1.49}$$

The procedure for construction of these matrices is as follows. The $(i-1)$th row of the matrix T_{j-1} and the ith row of the matrix T_j, $i = 2, \cdots, j+1, j = 1, \cdots, k$, always form two neighboring rows of a Pascal triangle with $T_0 = [1]$. For example, the first row of T_0, the second row of T_1, the third row of T_2, and the fourth row of T_3, etc. form a Pascal triangle. Similarly, the first row of T_1, the second of T_2, the third row of T_3, etc., or the first row of T_2 and the second row of T_3, etc. also form Pascal triangles. As far as the first row of each matrix is concerned, the lth element of the first row of the jth matrix is the lth element of the last row of the same matrix multiplied by $(-1)^{j+l+1}$.

Using the bilinear transformation we can write

$$\begin{bmatrix} s^{n-1} \\ \vdots \\ s \\ 1 \end{bmatrix} = \frac{1}{(1+z^{-1})^{n-1}} \begin{bmatrix} (1-z^{-1})^{n-1} \\ \vdots \\ (1+z^{-1})^{n-2}(1-z^{-1}) \\ (1+z^{-1})^{n-1} \end{bmatrix} \tag{1.50}$$

and

$$\begin{bmatrix} s^{n} \\ s^{n-1} \\ \vdots \\ s \\ 1 \end{bmatrix} = \frac{1}{(1+z^{-1})^{n}} \begin{bmatrix} (1-z^{-1})^{n} \\ (1+z^{-1})(1-z^{-1})^{n-1} \\ \vdots \\ (1+z^{-1})(1+z^{-1})^{n-2}(1-z^{-1}) \\ (1+z^{-1})(1+z^{-1})^{n-1} \end{bmatrix} \tag{1.51}$$

The comparison of (1.50) and (1.51), for $n = 1$ and $n = k$, yields

$$S' = T_k Z' \tag{1.52}$$

where the scaling factor $1/(1+z^{-1})^k$, which does not affect the transfer function H^{mn}, has been dropped. We see that both s to z^{-1} and z^{-1} to s transformations in (1.48) have the same form. Hence, similarly to (1.52), we can write

$$Z' = T_k S' \tag{1.53}$$

and we see that, instead of the inverse matrix T_k^{-1}, the matrix T_k can be used for the inverse z to s transformation. We find that

$$T_k T_k = 2^k U \tag{1.54}$$

where U is a unit matrix. Hence, the normalization factor of matrix T_k is $1/\sqrt{2^k}$. The transposition of (1.52) and (1.53) gives

$$S = ZT_k' \tag{1.55}$$

and

$$Z = ST_k' \tag{1.56}$$

which completes the proof of relations

$$\begin{aligned}
&P_h^{mn} = P^{mn}T_k,\ P^{mn} = P_h^{mn}T_k,\ Q_h = QT_k,\ Q = Q_hT_k\\
&A_h^{mn} = A^{mn}T_k,\ A^{mn} = A_h^{mn}T_k,\ B_h = BT_k,\ B = B_hT_k\\
&P^{mn} = T_k'A_h^{mn},\ A_h^{mn} = T_k'P^{mn},\ Q = T_k'B_h,\ B_h = T_k'Q\\
&A^{mn} = T_k'P_h^{mn},\ P_h^{mn} = T_k'A^{mn},\ B = T_k'Q_h,\ Q_h = T_k'B
\end{aligned} \tag{1.57}$$

shown in the scheme in Figure 1.4.

1.6 PROBLEMS

1. On the basis of (1.6) prove that the Fourier series (1.4) in the complex form is equivalent to the Fourier series in the trigonometric form:

$$x(t) = \frac{a_0}{2} + \sum_{n=1}^{\infty}\left[a_n \cos\left(\frac{2\pi}{T_p}nt\right) + b_n \sin\left(\frac{2\pi}{T_p}nt\right)\right] \tag{1.58}$$

2. On the basis of the definition (1.10), calculate the Laplace transform $X(s)$ of the function $x(t) = e^{\lambda t}$.
3. Calculate the Laplace transforms shown in Table 1.1 of the functions $e^{-\alpha t}$, $\sin \omega_o t$, $\cos \omega_o t$, $e^{-\alpha t} \sin\beta t$, and $e^{-\alpha t}\cos \beta t$.
   Hint: Use the result from the previous example, introducing $\lambda = -\alpha$, $\lambda = j\omega_o$, or $\lambda = -\alpha + j\beta$.
4. Calculate the Fourier series defined by (1.4), (1.5) of the periodic function

$$\sum_{k=-\infty}^{\infty} \delta(t - kT) \tag{1.59}$$

5. Prove the relation

$$\mathcal{F}\left\{\sum_{k=-\infty}^{\infty} \delta(t - kT)\right\} = \omega_s \sum_{m=-\infty}^{\infty} \delta(\omega - m\omega_s) \tag{1.60}$$

used in (1.21), which means that the Fourier transform of a sequence of impulses is also a sequence of impulses.

Hint: Use the inverse Fourier transform given in (1.7) and the result of the previous problem.

6. Prove the relation (1.31).
7. Choose appropriate relations from (1.57) and calculate $H_1'(z_1, s_2)$, $H_1''(s_1, z_2)$, and $H_2(z_1, z_2)$ for the transfer function

$$H(s_1, s_2) = \frac{[s_1\ 1]P[s_2^2\ s_2\ 1]'}{[s_1\ 1]Q[s_2^2\ s_2\ 1]'} \tag{1.61}$$

where

$$P = \begin{bmatrix} 0 & 0 & 1 \\ 1 & 0 & 0 \end{bmatrix}, \quad Q = \begin{bmatrix} 0 & 0 & 1 \\ 1 & 0 & 1 \end{bmatrix} \tag{1.62}$$

Note that the 2-D transfer function (1.61) is obtained from the transfer function of the first-order high-pass filter:

$$H_o(s) = \frac{s}{s+1} \tag{1.63}$$

after the substitution $s = s_1 + s_2^2$.

2

Properties of Basic CMOS Cells

CMOS VLSI circuits, introduced in the 1960s, have since become the dominant silicon technology. In comparison to other technologies, the main advantages of CMOS technology are

- A small number of fabrication processes
- High layout density
- Reduced power consumption

These advantages make CMOS circuits low cost and very attractive on the market.

This chapter describes the properties of MOS transistors used to realize all basic cells of integrated circuits. Operation, performance, and design of these cells are also presented in this chapter.

2.1 CMOS TRANSISTOR CHARACTERISTICS

MOS transistors are the simplest active element of integrated circuits. MOS transistors were invented by J. E. Lilienfeld (between 1925 and 1932 he got several patents in Canada and United States) and by O. Heil (the first English patent for such a transistor was issued in 1935). However, it was the rapid development of technology in the early 1960s that allowed implementation of MOS transistors and their use on a wide scale.

An *n*-channel MOS transistor is presented in Figure 2.1. Two n^+ regions obtained by diffusion in *p*-type silicon (substrate) are called drain and source. The polysilicon gate is isolated from the *p*-type silicon by a thin gate oxide layer with

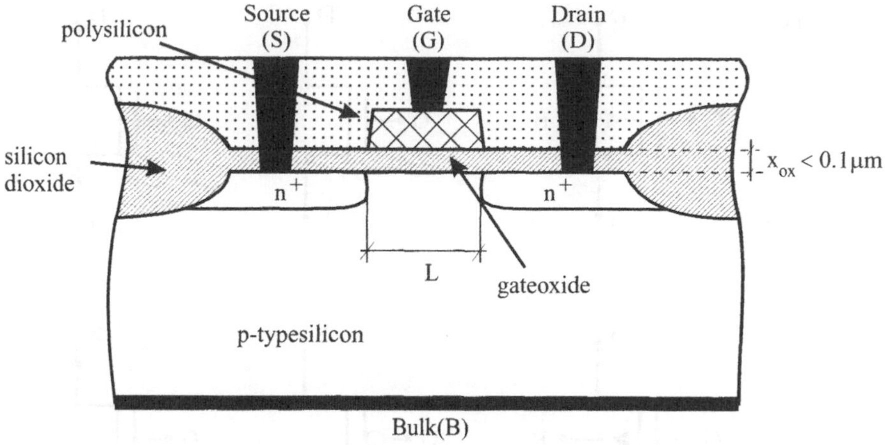

Figure 2.1 An *n*-channel MOS transistor.

the thickness x_{ox} less than a few hundredths of a micrometer. The process of obtaining such a device is described in the next chapter. Voltages are delivered to drain, source, and gate by metal (Al) contacts denoted as black boxes. The current flow I_D from drain to source is controlled by the drain-source voltage V_{DS}, the gate-source voltage V_{GS}, and the source-bulk voltage V_{SB}. Usually, there is a short circuit between the source and the bulk ($V_{SB} = 0$). In this case, the current I_D is controlled by voltages V_{DS} and V_{GS}, and the transistor can be in a cutoff mode ($I_D = 0$) or in an active mode ($I_D \neq 0$), depending on the parameter called the threshold voltage, V_{Tn}. When $V_{GS} \leq V_{Tn}$, the transistor is *off* (cutoff), and when $V_{GS} > V_{Tn}$, the transistor is *on* (active). In the cutoff mode, one of the *pn* junctions between substrate and drain or source is reverse-biased and the current I_D does not flow. In the active mode, the positive gate potential induces the conducting channel between drain and source. The channel is an electron surface with the length equal to L and the width equal to W. The section of a MOS transistor in Figure 2.1 shows only the channel length. The channel width is perpendicular to the figure plane. Because of the type of the channel, the transistor is called nMOS. The pMOS transistor is complementary to nMOS, which means that the current flows in the opposite direction and the voltages are reversed. The technology exploits the nMOS/pMOS pair of transistors is called CMOS technology. In Figure 2.2 different symbols of nMOS and pMOS transistors are presented. Let us note that we can use either complete symbols, which show four device electrodes, or simplified symbols, in which the bulk electrode is omitted. When the simplified symbols are used, the bulk electrode of an nMOS is assumed to be connected to the negative power supply (or to the ground) and that of a pMOS to the positive power supply.

Because the nMOS and pMOS transistors are complements of each other, we will examine the current–voltage characteristics of an nMOS transistor only. The

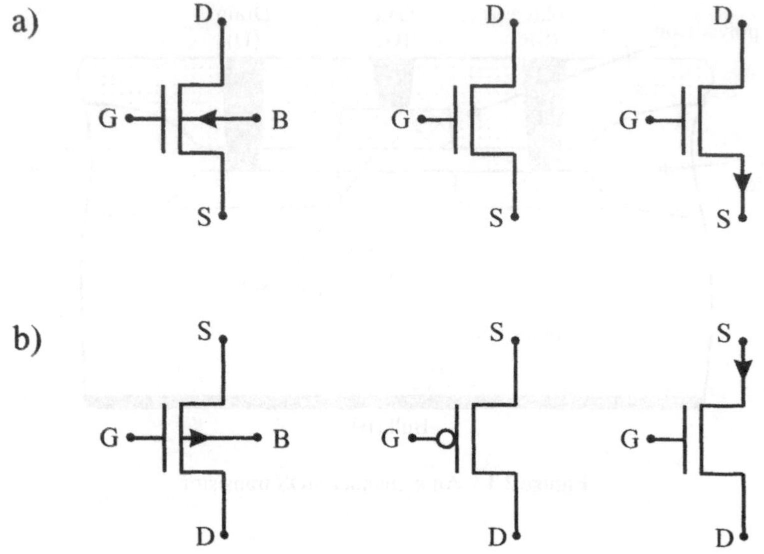

Figure 2.2 MOS transistor symbols: a) nMOS, b) pMOS.

simplest model is the square-low one, in which it is assumed that the drain current depends on drain-source voltage V_{DS} in the first and second power,

$$I_{Dn} = \frac{\beta_n}{2}[2(V_{GS} - V_{Tn})V_{DS} - V_{DS}^2] \tag{2.1}$$

exclusively. It is obvious that the device transconductance β_n is proportional to the width W and inversely proportional to the length L of the transistor channel and can be expressed as

$$\beta_n = k_n' \frac{W}{L} = \mu_n C_{ox} \frac{W}{L} \tag{2.2}$$

where k_n' is described by physical parameters:
μ_n = the electron mobility μ_p = hole mobility for pMOS transistors)
C_{ox} = the oxide capacitance per unit gate area

The unit area oxide capacitance can be calculated from

$$C_{ox} = \frac{\varepsilon_{ox}}{x_{ox}} \tag{2.3}$$

where $\varepsilon_{ox} = 3.9\varepsilon_0$, $\varepsilon_0 = (8.854E - 14)F/m$, and x_{ox} is the gate oxide thickness.

The coefficient of the first component in the formula (2.1) is linearly dependent on the gate-to-source voltage V_{GS}. For a given voltage V_{GS}, equation (2.1) describes the parabola shown in Figure 2.3. Only the part of the parabola between the origin of the coordinate system and the peak current of the parabola is used as the current–voltage characteristic. We say that when the values of current and voltages are in this region, the transistor is in a nonsaturated mode. It is easy to calculate from the extreme condition, $I'_{Dn}(V_{DS}) = 0$ that the saturation voltage has the value

$$V_{DS,\text{sat}} = V_{GS} - V_{Tn} \tag{2.4}$$

for which the value of current is

$$I_{Dn} = \frac{\beta_n}{2}(V_{GS} - V_{Tn})^2 \tag{2.5}$$

A more exact relation for the current in the saturated mode is in the form

$$I_{Dn} = \frac{\beta_n}{2}(V_{GS} - V_{Tn})^2[1 + \lambda_n (V_{DS} - V_{DS,\text{sat}})] \tag{2.6}$$

as shown in Figure 2.3, where λ_n is the n-channel-length modulation parameter. The dashed line in Figure 2.3 denotes the border between saturated and non-saturated areas.

The equation of a pMOS transistor in the nonsaturated mode can be written in the following form, complementary to (2.1):

$$I_{Dp} = \frac{\beta_p}{2}[2(V_{SG} + V_{Tp})V_{SD} - V_{SD}^2] \tag{2.7}$$

where the current is oriented as shown in Figure 2.2, the threshold voltage V_{Tp} now has a negative value, and $V_{GS} = -V_{SG} < V_{Tp}$. Similarly, in the saturated mode we have:

$$I_{Dp} = \frac{\beta_p}{2}(V_{SG} + V_{Tp})^2 \tag{2.8}$$

or

$$I_{Dp} = \frac{\beta_p}{2}(V_{SG} + V_{Tn})^2[1 + \lambda_p(V_{SD} - V_{SD,\text{sat}})] \tag{2.9}$$

where

$$V_{SD,\text{sat}} = V_{SG} + V_{Tp} \tag{2.10}$$

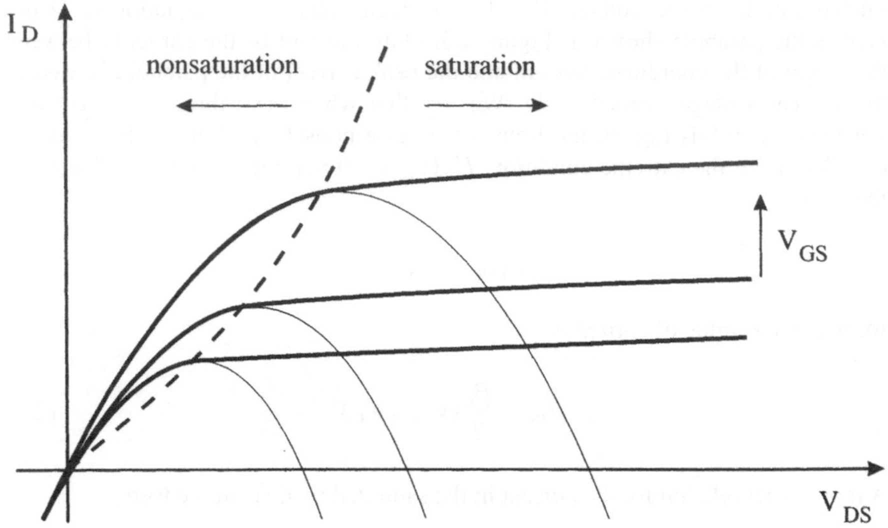

Figure 2.3 nMOS transistor current–voltage characteristics.

2.2 MODELING

Integrated circuit simulators, such as Spice, Eldo, or Saber, use more complex MOS transistor models than those that were described in the previous section. The parameters introduced so far, i.e.

- Transconductance parameter k'_n or k'_p
- Threshold voltages V_{Tn} and V_{Tp}
- Carrier mobility μ_n or μ_p
- Gate-oxide thickness x_{ox}
- Channel-length modulation parameter λ

are used on almost all levels of complexity of transistor models. The additional parameters that describe parasitic capacitors and parasitic resistors are very important, too. Parasitic capacitors of a MOS transistor are shown in Figure 2.4a. In order to calculate their capacitances, the technology parameters shown in Figure 2.4b are used at all levels of models. These parameters, and their units, are as follows:

- C_{GSO}—gate-source overlap capacitance/channel width, $[F/m]$
- C_{GDO}—gate-drain overlap capacitance/channel width, $[F/m]$
- C_{GBO}—gate-bulk overlap capacitance/channel length, $[F/m]$
- C_{j0}—zero-bias junction bottom capacitance/source (drain) area, $[F/m^2]$
- C_{jsw}— zero-bias junction sidewall capacitance/source (drain) perimeter length, $[F/m]$

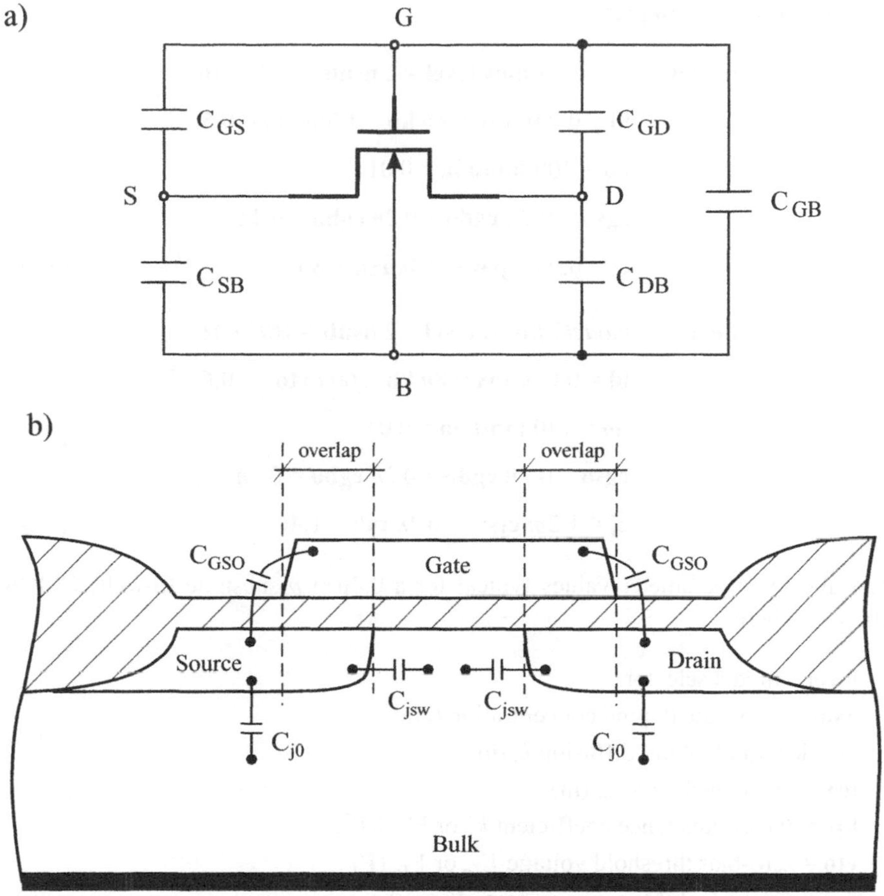

Figure 2.4 Parasitic capacitors (a), and technology parameters of a MOS transistor (b).

The gate-bulk overlap capacitance, C_{GBO}, cannot be shown in the cross-section presented in Figure 2.4.

The parameters are introduced with the use of a command that in Spice, for example, has the form

.model < *model name* > **nmos** [*model parameters*] (2.11)

for an NMOS transistor, or

.model < *model name* > **pmos** [*model parameters*] (2.12)

for a PMOS one.

The following examples

.model *model*1 **nmos level** = 2 **nsub** = $20E + 16$
+ **ld** = $0.05u$ **tox** = $8n$ **kp** = $170u$ **vto** = 0.5
+ **uo** = 400 **lambda** = 0.01
+ **cgso** = $0.2n$ **cgdo** = $0.2n$ **cgbo** = $0.1n$
+ **cj** = $0.9m$ **cjsw** = $0.3n$ **rsh** = 80 (2.13)

.model *model*2 **pmos level** = 2 **nsub** = $10E + 16$
+ **ld** = $0.05u$ **tox** = $8n$ **kp** = $60u$ **vto** = –0.6
+ **uo** = 140 **lambda** = 0.03
+ **cgso** = $0.2n$ **cgdo** = $0.2n$ **cgbo** = $0.1n$
+ **cj** = $1.2m$ **cjsw** = $0.4n$ **rsh** = 160 (2.14)

describe some parameter values typical for a 0.4μm *p*-substrate (*n*-well) CMOS process, where

level = model selector
nsub = substrate doping concentration (cm^{-3})
ld = length of lateral diffusion l_0 (m)
tox = oxide thickness x_{ox} (m)
kp = transconductance coefficient k_n' or k_p' (A/V^2)
vto = zero-bias threshold voltage V_{Tn} or V_{Tp} (V)
uo = surface mobility μ_n or μ_p ($cm^2/V \cdot sec$)
lambda = channel-length modulation parameter λ (V^{-1})
cgso = gate-source overlap capacitance C_{GSO} (F/m)
cgdo = gate-drain overlap capacitance C_{GDO} (F/m)
cgbo = gate-bulk overlap capacitance C_{GBO} (F/m)
cj = zero-bias junction bottom capacitance C_{j0} (F/m^2)
cjsw = zero-bias junction sidewall capacitance C_{jsw} (F/m)
rsh = drain or source diffusion sheet resistance R_{sh} (Ω/*square*)

In Spice, MOS transistor description has the general form

M < *name* > < *drain node* > < *gate node* > < *source node* >
+ < *bulk/substrate node* > < *model name* >
+ [**l** = < *value* >] [**w** = < *value* >]
+ [**ad** = < *value* >] [**as** = < *value* >]

+ **[pd** = < *value* >] **[ps** = < *value* >]
+ **[nrd** = < *value* >] **[nrs** = < *value* >]
+ **[nrg** = < *value* >] **[nrb** = < *value* >]
+ **[m** = < *value* >] (2.15)

where

l = channel length (m)
w = channel width (m)
ad = drain diffusion area (m^2)
as = source diffusion area (m^2)
pd = drain diffusion perimete (m)
ps = source diffusion perimeter (m)
nrd = relative drain resistivity (*squares*)
nrs = relative source resistivity (*squares*)
nrg = relative gate resistivity (*squares*)
nrb = relative substrate resistivity (*squares*)
m = device multiplier, simulating parallel devices

This description allows to calculate parasitic capacitors and resistors on the basis of given channel dimensions and drain and source areas and perimeters. The following examples

M*n*1 2 1 3 3 *model*1 **l** = 0.4*u* **w** = 1.2*u*
+ **ad** = 2.4*p* **as** = 2.4*p* **pd** = 6.4*u* **ps** = 6.4*u*
+ **nrd** = 10 **nrs** = 10 **nrg** = 8

and

M*p*1 2 1 4 4 *model2* **l** = 0.4*u* **w** = 3.6*u*
+ **ad** = 7.2*p* **as** = 7.2*p* **pd** = 11.2*u* **ps** = 11.2*u*
+ **nrd** = 30 **nrs** = 30 **nrg** = 24

illustrate the description (2.15) of nMOS and pMOS transistors, respectively. In these examples, the parameters from (2.13) and (2.14) are used.

2.3 SIMPLE CIRCUITS COMPOSED OF MOS TRANSISTORS

In this section, simple CMOS cells composed of one, two, or several transistors will be examined. All considerations will be based on the simplest equations given by relations (2.1) and (2.5) for an nMOS transistor and by (2.7) and (2.8) for a pMOS one. From these equations, we can obtain initial parameters of the designed cell, which can next be improved with the use of circuit simulators such as SPICE.

2.3.1 Voltage-Controlled Resistor

Transistor characteristics shown in Figure 2.3 allow us to consider a single transistor in a nonsaturated mode as a nonlinear resistor controlled by the voltage V_{GS}. From (2.1) we obtain the following expression for the conductance G_{tn}:

$$G_{tn} = \frac{I_{Dn}}{V_{DS}} = \beta_n(V_{GS} - V_{Tn}) - \frac{\beta_n}{2}V_{DS} \tag{2.16}$$

Equation (2.16) means that total conductance G_{tn} has two components corresponding to parallel resistors: a linear component $G_n = \beta_n(V_{GS} - V_{Tn})$ controlled by the voltage V_{GS} and a nonlinear negative (active) one $G_{na} = -\beta_n V_{DS}/2$. The linear voltage-controlled resistor has the resistance

$$R_n = \frac{1}{G_n} = \frac{1}{\beta_n(V_{GS} - V_{Tn})} \tag{2.17}$$

which can approximate the resitance of an nMOS transistor in the non-saturated mode during simplified analysis. For the pMOS transistor, from (2.7) we have the total conductance in the form

$$G_{tp} = \frac{I_{Dp}}{V_{SD}} = \beta_p(V_{SG} + V_{Tp}) - \frac{\beta_p}{2}V_{SD} \tag{2.18}$$

and the approximate linear value of the resistor is

$$R_p = \frac{1}{\beta_p(V_{SG} + V_{Tp})} \tag{2.19}$$

2.3.2 Diode-Connected Transistor

A MOS transistor with the gate and drain connected to each other is said to be diode connected. Let us consider a diode-connected nMOS transistor presented in Figure 2.5. For the connection of the gate shown in this figure, the transistor is *on* if $V > V_{Tn}$ and positive current flow I_{in} is possible. If $V \leq V_{Tn}$, the transistor is *off* and the current flow is impossible, which corresponds to a reversed-biased junction.

In Figure 2.6 two complementary pairs of diode-connected transistors are shown. Assuming that the transistors are in a saturated mode, we will show that the circuit in Figure 2.6a can be used as a resistor R, whereas the circuit in Figure 2.6b can be used as a voltage reference V_r. For the circuit in Figure 2.6a we have

$$I + \frac{\beta_p}{2}(V_{DD} - V + V_{Tp})^2 = \frac{\beta_n}{2}(V - V_{SS} - V_{Tn})^2 \tag{2.20}$$

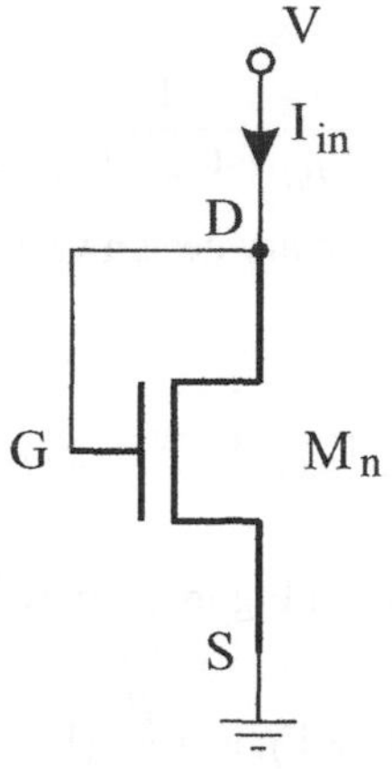

Figure 2.5 Diode-connected nMOS transistor.

and assuming

$$\beta_p = \beta_n = \beta \tag{2.21}$$

we have

$$I = \beta(V_{DD} + V_{Tp} - V_{SS} - V_{Tn})V + \frac{\beta}{2}[(V_{SS} + V_{Tn})^2 - (V_{DD} + V_{Tp})^2] \tag{2.22}$$

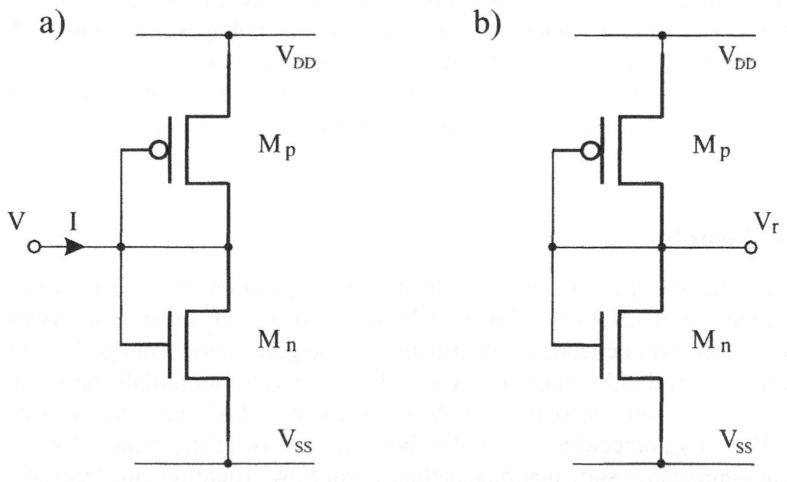

Figure 2.6 Diode-connected CMOS transistors.

Hence

$$R = \frac{1}{\beta(V_{DD} + V_{Tp} - V_{SS} - V_{Tn})} \tag{2.23}$$

where the parameters (2.21) are primary design variables. DC offset of the current I is given by the relation

$$I_{DC} = \frac{\beta}{2}[(V_{SS} + V_{Tn})^2 - (V_{DD} + V_{Tp})^2] \tag{2.24}$$

Equating currents in the circuit in Figure 2.6b, we get

$$\frac{\beta_p}{2}(V_{DD} - V_r + V_{Tp})^2 = \frac{\beta_n}{2}(V_r - V_{SS} - V_{Tn})^2 \tag{2.25}$$

and

$$V_r = \frac{\sqrt{\beta_p}(V_{DD} + V_{Tp}) + \sqrt{\beta_n}(V_{SS} + V_{Tn})}{\sqrt{\beta_n} + \sqrt{\beta_p}} \tag{2.26}$$

Equation (2.2) shows that the parameters β_n and β_p depend on the aspect ratio W/L. According to (2.26), channel widths and lengths can be used as adjusting parameters in order to obtain the required value of the reference voltage V_r in the range $[(V_{SS} + V_{Tn}), (V_{DD} + V_{Tp})]$.

2.3.3 Current Source

Considering equations (2.5) and (2.8), we see that nMOS and pMOS transistors in a saturated mode are ideal current sources with the current flow described by these equations. The current values depend on gate-source voltages. Equations (2.6) and (2.9) imply that in a more realistic description, the finite values of internal resistors in these sources should be taken into account. Improved performance can be obtained if two MOS transistors in cascode connection are introduced instead of a single one.

2.3.4 Switch

The MOS transistor can be in two basic modes of operation: in the active (on) mode and in the cutoff (off) one. Hence, a MOS transistor can serve as a switch. The mode of operation depends on the relation of the gate-source voltage V_{GS} with respect to the threshold voltage V_T. A switch composed of an nMOS transistor is *on* when $V_{GS} > V_{Tn}$, otherwise it is *off*. A switch composed of a pMOS transistor is *on* when $V_{GS} < V_{Tp}$; otherwise it is *off*. For both kinds of switches, in the *off* mode there is a high-impedance state that blocks the current flow. Therefore, the transistors can be considered as ideal switches in the *off* mode. The analysis is more complicated when the switch is on. In order to simplify the analysis, we will consider two cases of the transistor operating in the *on* mode:

1. The gate potential V_G is constant while the input voltage changes
2. The gate potential changes (the transistor is switched *on* or *off*) and the input voltage is constant during the switching process

In the second case, significant parasitic effects can occur.

First, let us consider a switch composed of an nMOS transistor whose input is connected to an impulse voltage source. The switch can be loaded with a resistor R or a capacitor C, as shown in Figure 2.7 a and b, respectively. In the *on* mode we place high voltage $V_G = V_{DD}$ at the gate. Let us consider the switch loaded with a resistor. On the rising slope of the positive input pulse, the transistor is in the nonsaturated mode and the output voltage rises. In this period V_{DS} increases and V_{GS} decreases. When V_{DS} achieves the saturation value, $V_{DS} = V_{DS,\text{sat}} = V_{GS} - V_{Tn}$, the switch transistor reaches the saturated mode in which the value of current I_{Dn} is determined by relation (2.5). The output voltage V_{out}, shown in Figure 2.8, has its maximum value $V_{\text{out}} = I_{Dn} \cdot R$ and can be described by the equations

$$I_{Dn}R = V_{\text{out}} = \frac{\beta_n R}{2}(V_{GS} - V_{Tn})^2,\ V_{GS} > V_{Tn} \tag{2.27}$$

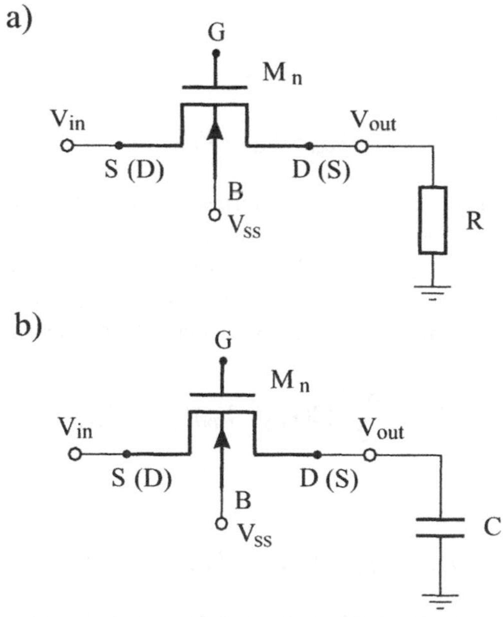

Figure 2.7 nMOS as a switch loaded by a resistor (a), or a capacitor (b).

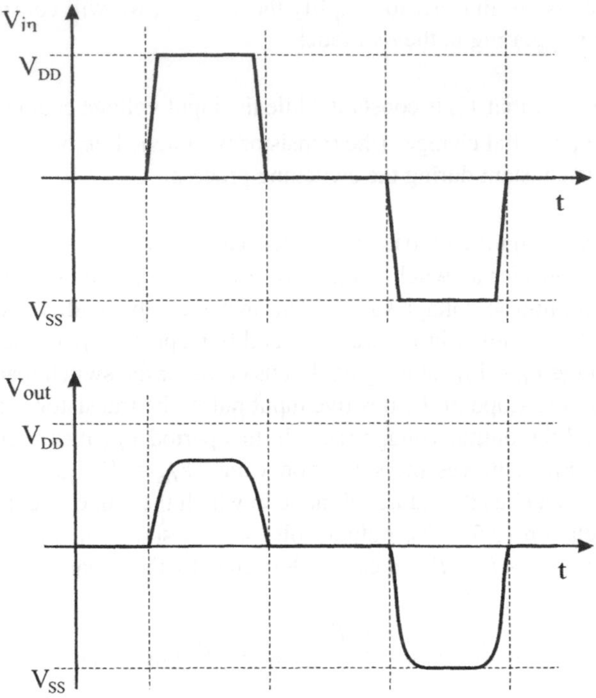

Figure 2.8 Transient characteristics of an nMOS switch loaded by a resistor.

and

$$V_{\text{out}} + V_{GS} = V_{DD} \tag{2.28}$$

We see that the transistor is in saturation during almost all time of excitation by the input source V_{in}. Similar analysis, with reversed roles of drain and source, can be performed for a negative input pulse. In this case, the transistor is in a nonsaturated mode for the whole input pulse duration. Hence, we use equation (2.1) of the transistor in the nonsaturated mode in order to calculate the maximum value of $|V_{\text{out}}|$:

$$V_{\text{out}} = -I_{Dn}R = -\frac{\beta_n R}{2}[2(V_{GS} - V_{Tn})V_{DS} - V_{DS}^2],\ V_{GS} > V_{Tn} \tag{2.29}$$

where

$$V_{GS} = V_{DD} - V_{SS}, \qquad V_{DS} = V_{\text{out}} - V_{SS} \tag{2.30}$$

A different situation occurs when the switch is loaded with a capacitor. In this case, the output voltage is described by the equation

$$I_{Dn} = C\frac{dV_{\text{out}}}{dt} \tag{2.31}$$

At the beginning of the excitation period, the transistor is in a nonsaturated mode. However, after a very short time the input impulse achieves its maximum value and the transistor changes into the saturated mode. Hence, in order to determine the charging time constant we can use relations (2.5) and (2.31), which yield

$$C\frac{dV_{\text{out}}}{dt} = \frac{\beta_n}{2}(V_{DD} - V_{\text{out}} - V_{Tn})^2 \tag{2.32}$$

It is a special case of the Riccati equation

$$y' = a(t)y^2 + b(t)y + c(t) \tag{2.33}$$

which can be solved with the use of substitution

$$V_{\text{out}} = V_{DD} - V_{Tn} + \frac{1}{z} \tag{2.34}$$

The solution which we obtain is in the form

$$V_{\text{out}} = V_{DD} - V_{Tn} + \frac{1}{A - \dfrac{\beta_n}{2C}t} \tag{2.35}$$

with the integral constant A calculated from the initial condition $V_{\text{out}}(0) = 0$. Hence, the capacitor load characteristic is described by the relation

$$V_{\text{out}} = V_{DD} - V_{Tn} - \frac{V_{DD} - V_{Tn}}{\dfrac{t}{\tau_{ch}} + 1} = (V_{DD} - V_{Tn})\frac{\dfrac{t}{\tau_{ch}}}{\dfrac{t}{\tau_{ch}} + 1} \tag{2.36}$$

where

$$\tau_{ch} = \frac{2C}{\beta_n(V_{DD} - V_{Tn})} \tag{2.37}$$

In the discharge process, the current flows in the opposite direction, and the transistor, with interchanged roles of drain and source, is in the non-saturated mode in which its resistance can be estimated by the formula (2.17) with $V_{GS} = V_{DD}$. Hence,

$$\tau_{ch} = \frac{2C}{\beta_n(V_{DD} - V_{Tn})} > \tau_{dis} = \frac{C}{\beta_n(V_{DD} - V_{Tn})} \tag{2.38}$$

as shown in Figure 2.9.

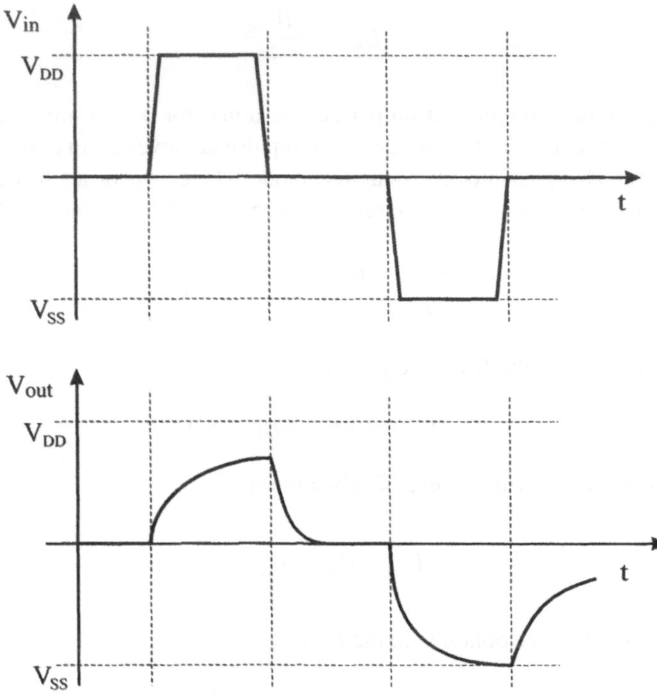

Figure 2.9 Transient characteristics of an nMOS switch loaded by a capacitor.

The behavior of the switch composed of a pMOS transistor is complementary to a switch composed of an nMOS one. Hence, the switch composed of a complementary transistor pair, presented in Figure 2.10, will have better symmetry properties. The transient characteristics of a switch composed of a complementary pair of MOS transistors is shown in Figure 2.11.

Two significant parasitic effects that have to be considered in CMOS switch operation are the clock-feed-through effect and the charge injection. These parasitic effects appear when a switch changes its mode of operation from *on* to *off* and backward. Let us assume that the switch is controlled by the clock shown in Figure 2.12a,b. The signal Φ_1 is delivered to the gate of the nMOS transistor, whereas the signal Φ_2 is delivered to the gate of the pMOS transistor. Hence, the switch is *on* in the odd clock period and *off* in the even one. A so-called nonoverlapping clock (pulses of the clocks Φ_1 and Φ_2 do not overlap) with a 50% filling coefficient is used. Because of finite rise and fall times of signals, the filling coefficient is usually less than 50% in order to provide a guard interval for the proper operation of different switches in the circuit. The voltage delivered to the input of the switch is shown in Figure 2.12c, whereas the voltage on the load capacitance is shown in Figure 2.12d. When the switch is *on*, V_{out} changes exponentially, as was explained previously. In the *off* mode, the output signal is held. However, its value is not exactly

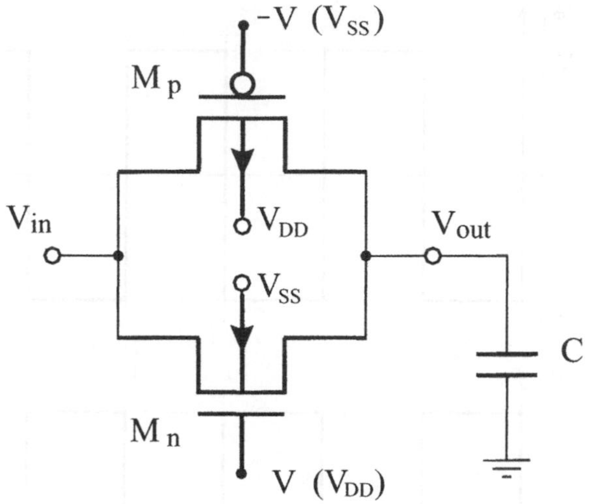

Figure 2.10 Switch composed of CMOS transistors.

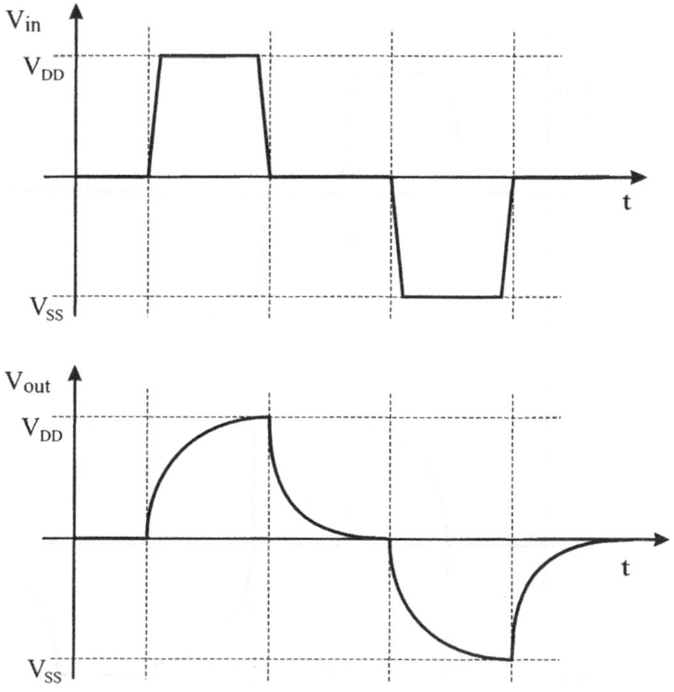

Figure 2.11 Transient characteristics of a switch composed of CMOS transistors.

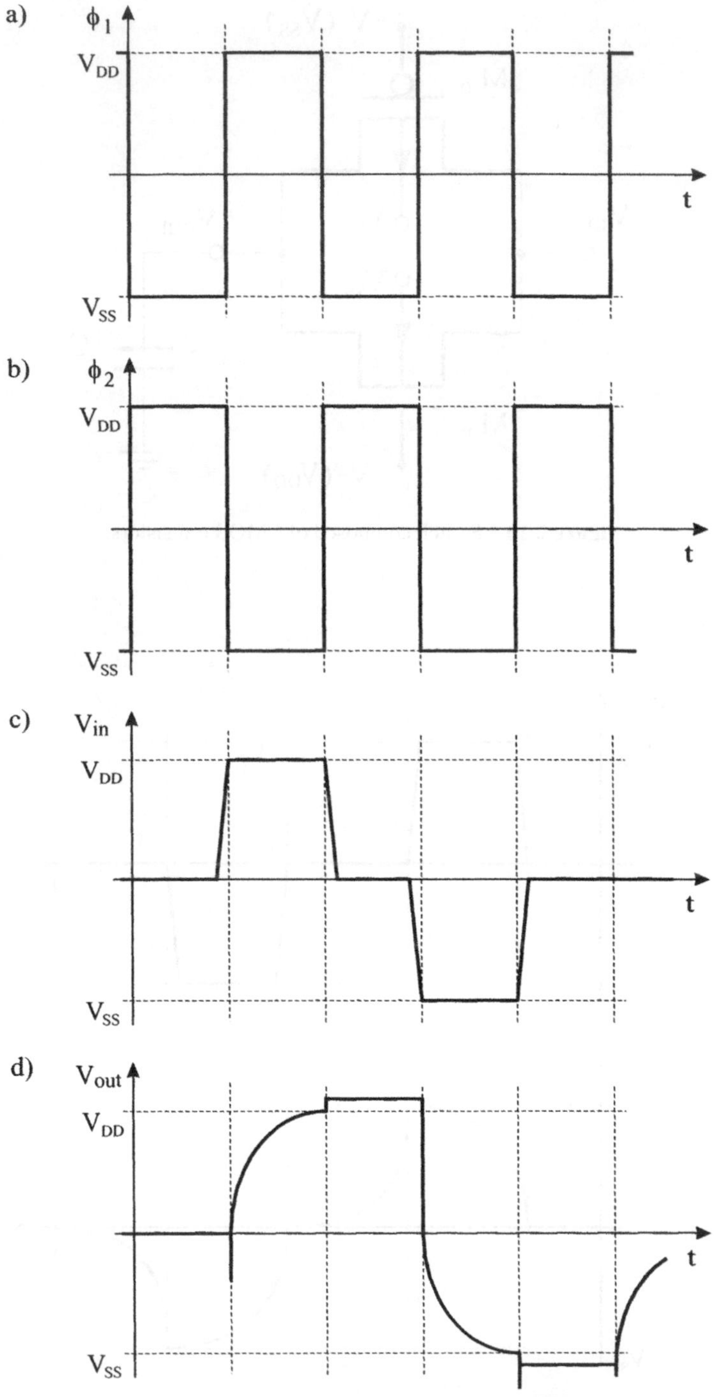

Figure 2.12 Parasitic effects in a switch composed of MOS transistors.

the same as before the switching, as part of the charge from the transistor channels is injected into the output capacitance and not into the bulk. Glitches shown in Figure 2.12d are caused by gate capacitances, which play the role of differentiators of the clock signals. The shorter rise and fall times, the greater these parasitic effects.

2.3.4.1 *CMOS Switch Logic* The CMOS switch shown in Figure 2.9 can serve as a logic transmission gate (*TG*). Logical values 0 and 1 are associated with the low (zero) and high (V_{DD}) voltages, will be described in detail in the next section. The symbol and the truth table of this logic device is shown in Figure 2.13. The logical value from the input x_1 is transferred to the output y when $x_2 = 1$. When $x_2 = 0$, the *TG* is off and the output signal can be memorized on the output parasitic capacitance. However, in real circuits the leakage currents connect the output node to the ground and $y = 0$. Hence, the transmission gate realizes the *AND* logic function of the input variables x_1 and x_2.

The realization of the *XOR* function

$$y = F(x_1, x_2) = x_1\bar{x}_2 + \bar{x}_1 x_2 = x_1 \oplus x_2 \tag{2.39}$$

is shown in Figure 2.14a. An equivalent *XOR* gate in the form of a split array, which is more convenient for layout design, is shown in Figure 2.14b.

2.3.5 Inverter

The circuit shown in Figure 2.15, composed of a complementary pair of MOS transistors, is called an inverter. In order to examine its DC transfer curve, we assume that the input voltage changes in the range (V_{SS}, V_{DD}). The DC characteristic of the inverter is presented in Figure 2.16. If $V_{\text{out}} = (V_{DD} - V_{SS})/2$ for $V_{\text{in}} = 0$, then the inverter is called symmetrical. It is possible to obtain the symmetrical inverter if the fabrication process is polarity-symmetric with

$$V_{Tn} = -V_{Tp} \tag{2.40}$$

and if the inverter is designed for

$$\beta_n = \beta_p \tag{2.41}$$

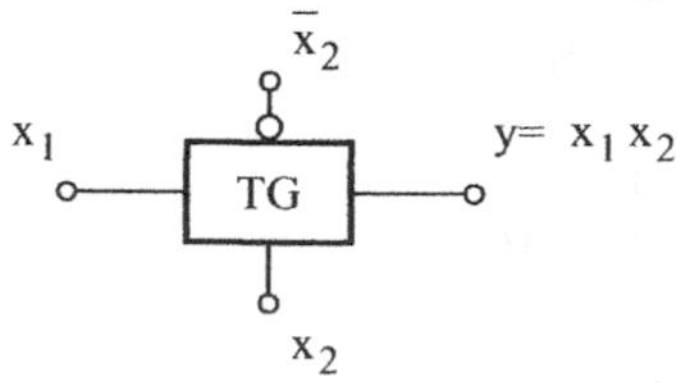

x_1	x_2	y
0	0	0
1	0	0
0	1	0
1	1	1

Figure 2.13 Symbol and truth table of a transmission gate composed of a CMOS switch.

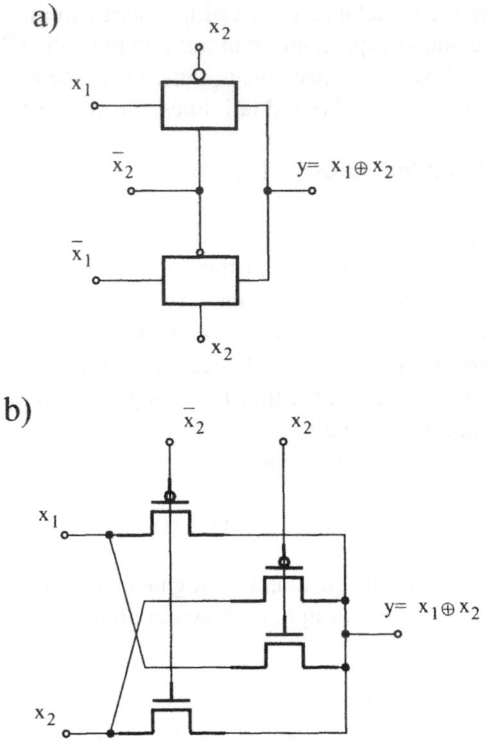

Figure 2.14 XOR gate based on a transmission gate (a) and on a split array (b).

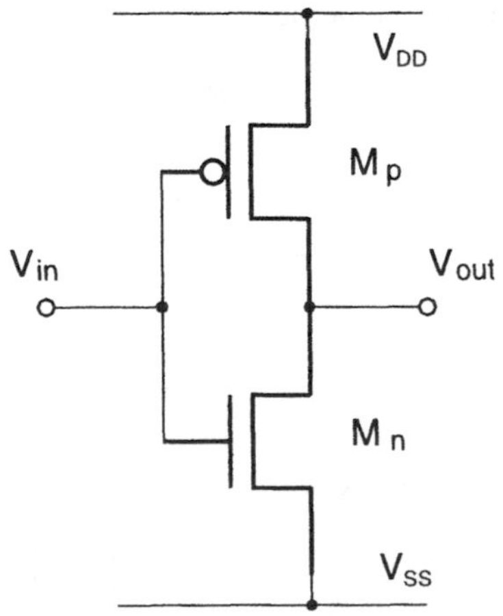

Figure 2.15 CMOS inverter.

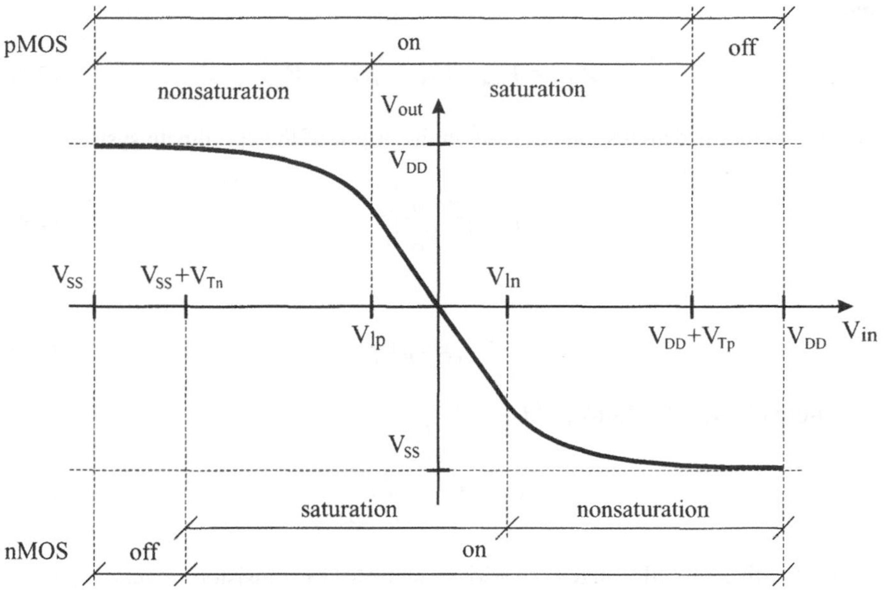

Figure 2.16 DC transfer curve of an inverter.

For the voltages given in Figure 2.15 we have

$$V_{GSn} = V_{\text{in}} - V_{SS}, \qquad V_{DSn} = V_{\text{out}} - V_{SS} \tag{2.42}$$

for the nMOS transistor and

$$V_{SGp} = V_{DD} - V_{\text{in}}, \qquad V_{SDp} = V_{DD} - V_{\text{out}} \tag{2.43}$$

for the pMOS. In the inverter characteristic, five ranges of voltage V_{in} in the inverter operation are shown. In the first range, the nMOS transistor is *off*, while the pMOS one is conducting in the nonsaturated mode. Hence, V_{out} reaches its maximum value V_{DD}. In the second range, the pMOS transistor is in the nonsaturated mode until V_{in} achieves a limit value V_{Ip}. However, the nMOS transistor is conducting in the saturated mode, which gives a nonlinear decrease of the output voltage V_{out}. In the third range, both transistors are conducting in the saturated mode. This range is located between the points V_{Ip} and V_{In}, which can be calculated on the basis of the saturation voltages (2.4) and (2.10). For example, saturation of an nMOS transistor occurs at the point

$$V_{DSn} = V_{DSn,\text{sat}} \tag{2.44}$$

where

$$V_{DSn,\text{sat}} = V_{SGn} - V_{Tn} = V_{\text{in}} - V_{SS} - V_{Tn} \tag{2.45}$$

and

$$V_{DSn} = V_{\text{out}} - V_{SS} = -\mu V_{\text{in}} - V_{SS} \tag{2.46}$$

The coefficient μ denotes voltage gain in the origin of the coordinate system

$$\mu = \left| \frac{dV_{\text{out}}}{dV_{\text{in}}} \right|_{V_{\text{in}}=0} \tag{2.47}$$

or

$$V_{\text{out}} = -\mu V_{\text{in}} \tag{2.48}$$

Equations (2.44, 2.45, 2.46) yield

$$V_{\text{in}} = V_{In} = \frac{V_{Tn}}{1+\mu} \tag{2.49}$$

In the fourth and fifth ranges the nMOS and pMOS transistors reverse operation modes compared to the second and first range, respectively.

Concluding, the most important relations describing the five regions of the DC transfer curve are as follows:

1. $V_{\text{in}} \in (V_{SS}, V_{SS} + V_{Tn})$: $V_{GSn} < V_{Tn}$, $V_{SDp} = 0 \leq V_{SDP,\text{sat}}$
2. $V_{\text{in}} \in (V_{SS} + V_{Tn}, V_{Ip})$: $V_{GSn} > V_{Tn}$, $V_{SDp} \leq V_{SDP,\text{sat}}$
3. $V_{\text{in}} \in (V_{Ip}, V_{In})$: $V_{DSn} > V_{DSn,\text{sat}}$, $V_{SDp} > V_{SDP,\text{sat}}$
4. $V_{\text{in}} \in (V_{In}, V_{DD} + V_{Tp})$: $V_{SGp} > -V_{Tp}$, $V_{DSn} \leq V_{DSn,\text{sat}}$
5. $V_{\text{in}} \in (V_{DD} + V_{Tp}, V_{DD})$: $V_{SGp} < -V_{Tn}$, $V_{DSn} = 0 \leq V_{DSn,\text{sat}}$

2.3.5.1 *Inverter as a Basic Amplifier* In the third range of the DC transfer curve, in which both transistors are saturated, the inverter can be considered as a transconductance amplifier. The output current of the inverter is given by the equation

$$I_{\text{out}} = I_{Dp} - I_{Dn} = \frac{\beta_p}{2}(V_{SGp} + V_{Tp})^2 - \frac{\beta_n}{2}(V_{GSn} - V_{Tn})^2 \tag{2.50}$$

Hence, for the gate voltages given by (2.42, 2.43) we obtain

$$\begin{aligned} I_{\text{out}} = {} & (\beta_p - \beta_n)V_{\text{in}}^2/2 - [\beta_p(V_{DD} + V_{Tp}) - \beta_n(V_{SS} + V_{Tn})]V_{\text{in}} \\ & + \beta_p(V_{DD} + V_{Tp})^2/2 - \beta_n(V_{SS} + V_{Tn})^2/2 \end{aligned} \tag{2.51}$$

For $\beta_p = \beta_n$ the first component is cancelled and the output current I_{out} depends linearly on the input voltage V_{in}. The third component denoting DC offset of the output current can also be cancelled for a symmetrical inverter $V_{Tp} = V_{Tn}$ and a symmetrical power supply $V_{DD} = -V_{SS}$.

The output voltage depends on the load. If we assume that the amplifier is not loaded, the output voltage is given by

$$V_{\text{out}} = I_{\text{out}} R_o \tag{2.52}$$

where R_o is the total drain-to-source resistance. Hence, the low-frequency, small-signal voltage gain is calculated from

$$\mu = -\frac{V_{\text{out}}}{V_{\text{in}}} = -\frac{I_{\text{out}} R_o}{V_{\text{in}}} = \frac{(g_n + g_p) V_{\text{in}}}{V_{\text{in}}(g_{dsn} + g_{dsp})} = \frac{g_n + g_p}{g_{dsn} + g_{dsp}} \tag{2.53}$$

where g_n and g_p are trasconductances and g_{dsn} and g_{dsp} are drain-to-source conductances in small-signal models of nMOS and pMOS transistors, respectively.

2.3.5.2 *Inverter as a Digital Logic Circuit* On the assumption that $V_{SS} = 0$, the DC transfer curve of the inverter has the form shown in Figure 2.17. Let us associate the logic values "0" and "1" with the voltage ranges in the following way:

$$V_{\text{in}} \in <0, V_{IL}> \doteq \text{"0"},\; V_{\text{in}} \in <V_{IH}, V_{DD}> \doteq \text{"1"} \tag{2.54}$$

and

$$V_{\text{out}} \in <0, V_{OL}> \doteq \text{"0"},\; V_{\text{out}} \in <V_{OH}, V_{DD}> \doteq \text{"1"} \tag{2.55}$$

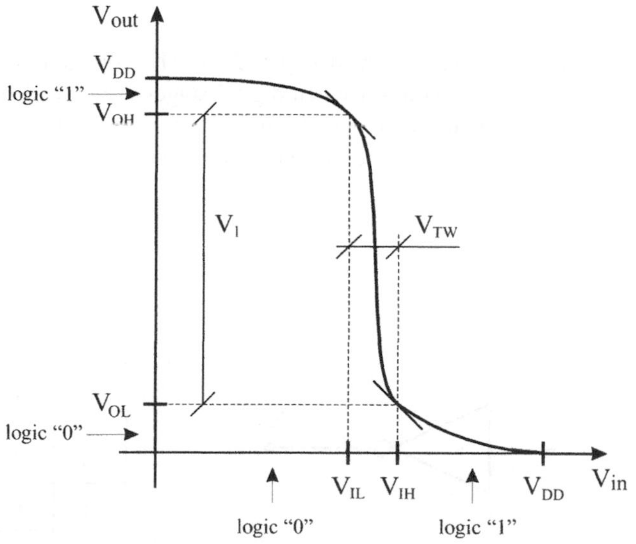

Figure 2.17 Digital inverter characteristics.

The voltages V_{IL}, V_{OH}, and V_{IH}, V_{OL} determine the coordinates of the points where the slope of the digital inverter characteristic is –1, i.e.,

$$\frac{dV_{\text{out}}}{dV_{\text{in}}} = -1 \tag{2.56}$$

In the regions defined by equations (2.54), (2.55), and (2.56), logic gates work on condition that

$$\left|\frac{dV_{\text{out}}}{dV_{\text{in}}}\right| < 1 \tag{2.57}$$

and they attenuate the noise added to their input.

We see that for the associations (2.54) and (2.55), the inverter works as a digital element whose symbol and truth table are shown in Figure 2.18.

Apart from V_{IL}, V_{IH}, V_{OL}, and V_{OH} several other voltages are introduced in order to characterize a digital circuit:

1. The logic swing V_l is defined as $V_l = V_{OH} - V_{OL}$ and describes the maximum output voltage variation.
2. The transition width $V_{TW} = V_{IH} - V_{IL}$ is a measure of input sensitivity.
3. The voltage noise margin is introduced both for high and low logic states, as $V_{NML} = V_{IL} - V_{OL}$, $V_{NMH} = V_{OH} - V_{IH}$.
4. The gate threshold voltage V_{th} is defined at the intersection of the inverter characteristic in Figure 2.17 and the unity gain line as the point where $V_{\text{in}} = V_{\text{out}} = V_{th}$.

The voltages V_l and V_{TW} are shown in Figure 2.17. The voltage noise margins denote the level of voltage separation when digital stages are cascaded. The voltage noise margins must be greater than zero for proper operation of the circuit.

The voltages V_{IL} and V_{IH} can be computed by combining the condition (2.56) with the equation

$$I_{Dn} = I_{Dp} \tag{2.58}$$

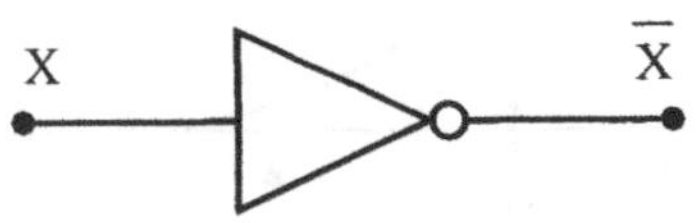

X	$\overline{X}$
0	1
1	0

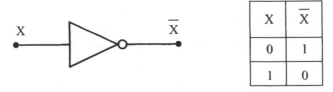

Figure 2.18 Symbol and truth table of a CMOS inverter.

V_{IL} is in the region in which the nMOS transistor is saturated and the pMOS transistor is nonsaturated. Hence, I_{Dn} and I_{Dp} are determined by (2.5) and (2.7), respectively, and the equations (2.58) and (2.56) can be written in the form

$$\beta_n(V_{\text{in}} - V_{Tn})^2 = \beta_p[2(V_{DD} - V_{\text{in}} + V_{Tp}) - (V_{DD} - V_{\text{out}})](V_{DD} - V_{\text{out}}) \quad (2.59)$$

and

$$\beta_n(V_{\text{in}} - V_{Tn}) = \beta_p(V_{DD} - V_{\text{in}} + V_{Tp}) - 2\beta_p(V_{DD} - V_{\text{out}}) \quad (2.60)$$

From the above equations we obtain $V_{IL} = V_{\text{in}}$ by eliminating V_{out}. In the region in which V_{IH} is determined, the nMOS transistor is nonsaturated and I_{Dn} is obtained from (2.1), whereas the pMOS transistor is saturated and I_{Dp} is obtained from (2.8). The equations from which $V_{IH} = V_{\text{in}}$ can be calculated have the form

$$\beta_n[2(V_{\text{in}} - V_{Tn}) - V_{\text{out}}]V_{\text{out}} = \beta_p(V_{DD} - V_{\text{in}} + V_{Tp})^2 \quad (2.61)$$

and

$$2\beta_n V_{\text{out}} - \beta_n(V_{\text{in}} - V_{Tn}) = -\beta_p V_{DD} + \beta_p(V_{\text{in}} - V_{Tp}) \quad (2.62)$$

The definition of the threshold voltage V_{th} given above means that it can be calculated from the relation

$$\beta_n(V_{th} - V_{Tn})^2 = \beta_p(V_{DD} - V_{th} + V_{Tp})^2 \quad (2.63)$$

obtained after equating drain currents of nMOS and pMOS transistors, respectively. Hence, V_{th} obtained from (2.63) has the form

$$V_{th} = \frac{\sqrt{\gamma}V_{Tn} + (V_{DD} + V_{Tp})}{1 + \sqrt{\gamma}} \quad (2.64)$$

where

$$\gamma = \frac{\beta_n}{\beta_p} \quad (2.65)$$

The threshold voltage is used to explain the operation of the Schmitt trigger, composed of two inverters and shown in Figure 2.19. For the input and output ports denoted in Figure 2.19, the first inverter, composed of transistors Mp_1 and Mn_1, is in the feedback loop of the second inverter composed of Mp_2 and Mn_2. Suppose that the input voltage V_{in}, applied to the input of the second inverter, is the impulse shown in Figure 2.20a. Because of the feedback capacitance, composed of gate capacitances of the first inverter transistors, the output impulse is delayed in comparison to the input impulse, as shown in Figure 2.20a. We can observe that the threshold voltage V_{thF} is greater on the rising slope than the threshold voltage V_{thR} on the falling slope. Hence, the trigger has a hysteresis loop, shown in Figure 2.20b, with

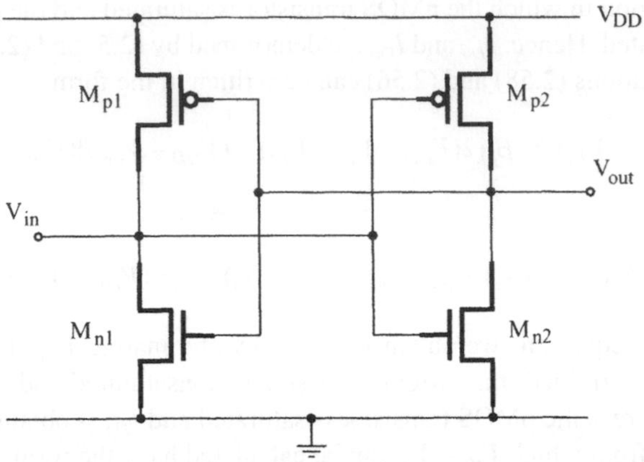

Figure 2.19 Schmitt trigger composed of two CMOS inverters.

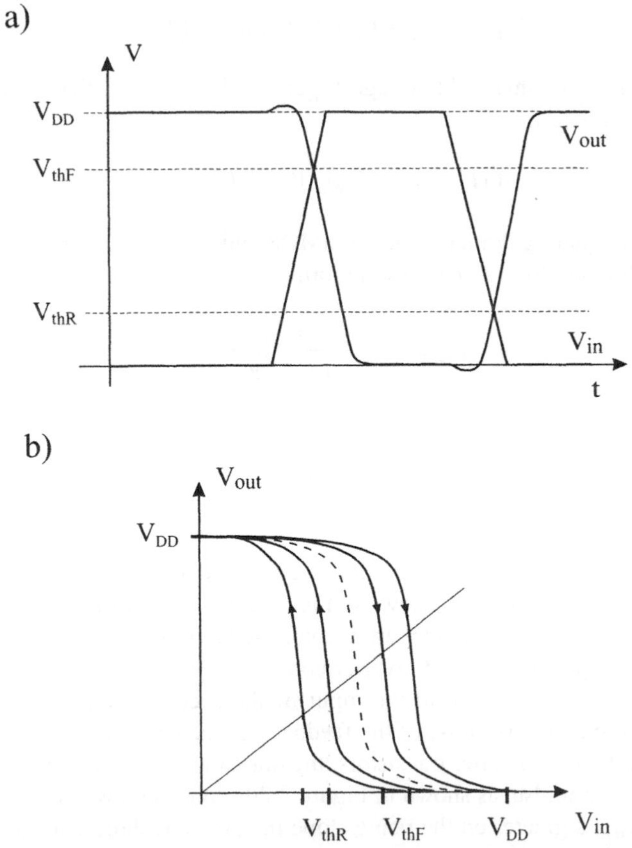

Figure 2.20 Input and output waveforms (a) and hysteresis (b) of Schmitt trigger.

V_{th} in the forward (F) switching event greater than V_{th} in the reverse (R) one. For steeper slopes, the hysteresis loop is wider.

Another interesting application of inverters is a D flip-flop (*DFF*), shown in Figure 2.21, controlled by the clock signals Φ and $\overline{\Phi}$ delivered to transmission gates (*TG*). In the clock phase $\Phi = 0$, the data *D* from the input is delivered to the input of the first inverter in the data path. In the clock phase $\Phi = 1$, this value is held by the inverter in the feedback branch of the first loop. In this clock phase, the data *D* is delivered from the output of the first inverter to the input of the second inverter in the data path and held in the same manner as in the first loop. Hence, the data is transferred from the input to the output with the delay equal to one clock period. Using *NAND* gates instead of inverters in the *DFF*s allows one to introduce clear and set inputs in these *DFF*s.

Transient characteristics determine the speed of digital circuit operation. Let us consider the inverter terminated with capacitance C_{out}, presented in Figure 2.22. The output capacitance includes several components. The most important components are drain-bulk parasitic capacitances C_{DBn} and C_{DBp} and gate-drain parasitic capacitances C_{GDn} and C_{GDp} of the complementary transistors and the capacitance of an interconnection with the next logic stage. The parasitic capacitances of both stages should be taken into account. The simple model of the inverter, which can be used in order to estimate the switching time, is presented in Figure 2.22b. We see that transistors play the role of switches and, depending on the input signal, C_{out} is charged through R_p or discharged through R_n. It follows from the considerations presented in the section concerning switches that the formulas (2.17) and (2.19) can be used to calculate these resistances, with the gate-source voltages equal to V_{DD}. Hence, the discharge time constant is given by

$$\tau_n = R_n C_{out} \tag{2.66}$$

where

$$R_n = \frac{1}{\beta_n(V_{DD} - V_{Tn})} \tag{2.67}$$

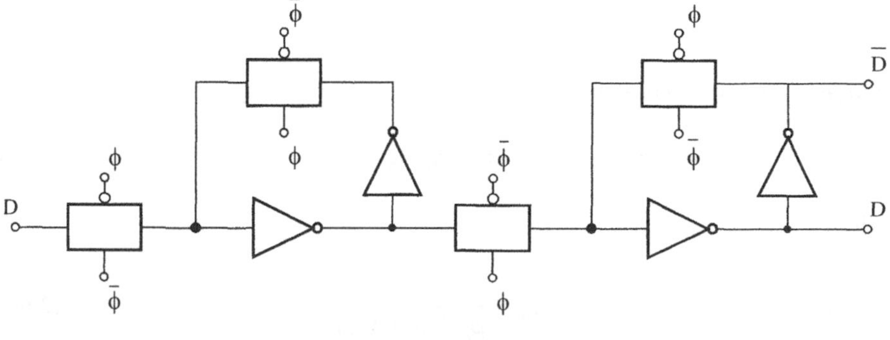

Figure 2.21 D flip-flop composed of inverters and transmission gates.

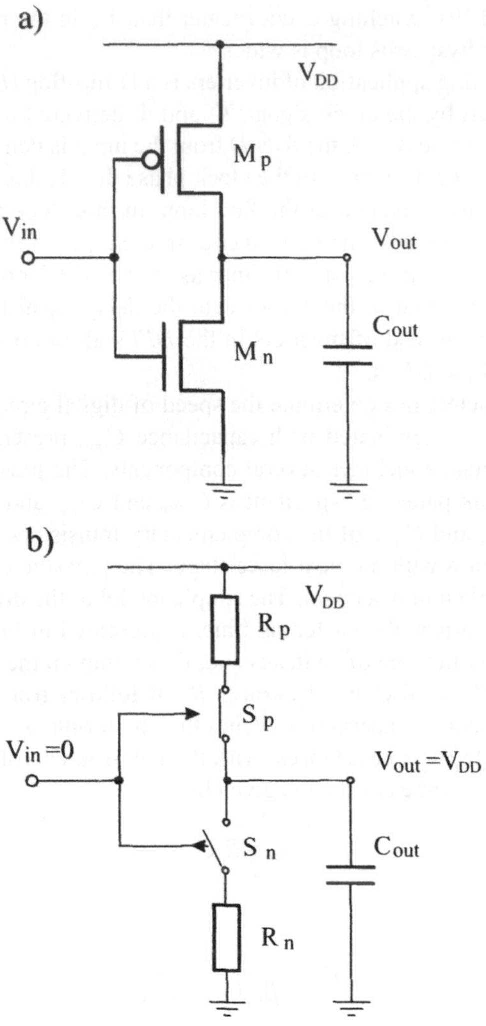

Figure 2.22 Inverter terminated by capacitance (a) and its transient switch model (b).

and the charge time constant by

$$\tau_p = R_p C_{\text{out}} \tag{2.68}$$

where

$$R_p = \frac{1}{\beta_p (V_{DD} + V_{Tp})} \tag{2.69}$$

From these formulas, for desired time constants we get the following aspect ratios W/L of the transistors:

$$\left(\frac{W}{L}\right)_n = \frac{C_{\text{out}}}{k'_n \tau_n (V_{DD} - V_{Tn})} \tag{2.70}$$

and

$$\left(\frac{W}{L}\right)_p = \frac{C_{\text{out}}}{k'_p \tau_p (V_{DD} + V_{Tp})} \tag{2.71}$$

Assuming that the high-to-low time is estimated by $t_{HL} \approx 3\tau_n$ and the low-to-high time by $t_{LH} \approx 3\tau_p$, we define the maximum switching frequency as

$$f_{\max} = \frac{1}{t_{HL} + t_{LH}} \tag{2.72}$$

Let us note that real logic gates are usually much slower in comparison with an estimation given by (2.72).

2.3.6 Current Mirror

A simple realization of a current mirror composed of two nMOS transistors is presented in Fig. 2.23. The transistor $M0$ is in diode connection, whereas the gate of the transistor $M1$ is connected with the gate of $M0$ and $V_{GS0} = V_{GS1} = V_{GS}$. The coefficient α_1 is determined by aspect ratios of both transistors

$$\alpha_1 = \frac{W_1/L_1}{W_0/L_0} \tag{2.73}$$

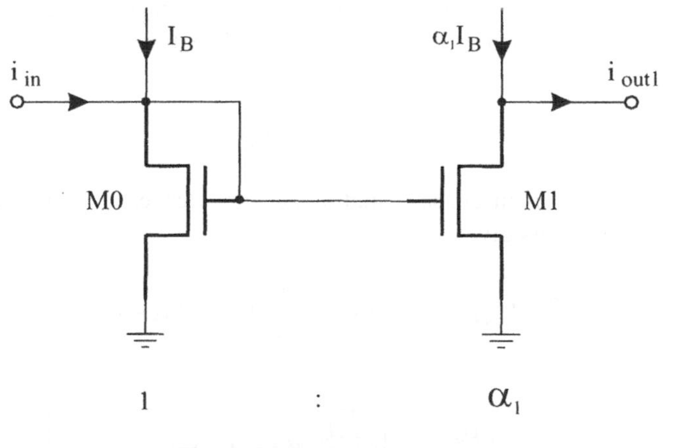

Figure 2.23 Current mirror.

Assuming that the bias current I_B keeps both transistors in a saturated mode, we have

$$\frac{\alpha_1 I_B - i_{\text{out}1}}{I_B + i_{\text{in}}} = \frac{\beta_1 (V_{GS} - V_{Tn})^2}{\beta_0 (V_{GS} - V_{Tn})^2} \tag{2.74}$$

or

$$\frac{\alpha_1 I_B - i_{\text{out}1}}{I_B + i_{\text{in}}} = \frac{W_1/L_1}{W_0/L_0} = \alpha_1 \tag{2.75}$$

and

$$i_{\text{out}1} = -\alpha_1 i_{\text{in}} \tag{2.76}$$

If $\alpha_1 = 1$ then the circuit mirrors the input current at its output.

Current mirrors with multiple outputs can be obtained by adding transistors $M2$ ... Mn connected to $M0$ in the same manner as $M1$ and described by the coefficients

$$\alpha_i = \frac{W_i/L_i}{W_0/L_0}, \; i = 2, \ldots, n \tag{2.77}$$

In this case we have

$$i_{\text{out}k} = -\alpha_k i_{\text{in}}, \; k = 2, \ldots, n \tag{2.78}$$

2.3.7 Amplifier Stage

An amplifier stage composed of a pair of nMOS and pMOS transistors is shown in Figure 2.24. The pMOS transistor is diode connected. The circuit in which the gate of the pMOS transistor is connected to the reference voltage source V_B is also often used. In the circuit shown in Figure 2.24, the pMOS transistor is in the saturated mode. If

$$V_{\text{out}} > V_{\text{in}} - V_{Tn} \tag{2.79}$$

the nMOS transistor is in the saturated mode, too. Hence, equating the drain currents of both transistors gives

$$\frac{\beta_p}{2}(V_{DD} - V_{\text{out}} + V_{Tp})^2 = \frac{\beta_n}{2}(V_{\text{in}} - V_{SS} - V_{Tn})^2 \tag{2.80}$$

and

$$V_{\text{out}} = -\sqrt{\frac{\beta_n}{\beta_p}}\, V_{\text{in}} + \left[\sqrt{\frac{\beta_n}{\beta_p}}(V_{SS} + V_{Tn}) + V_{DD} + V_{Tp}\right] \tag{2.81}$$

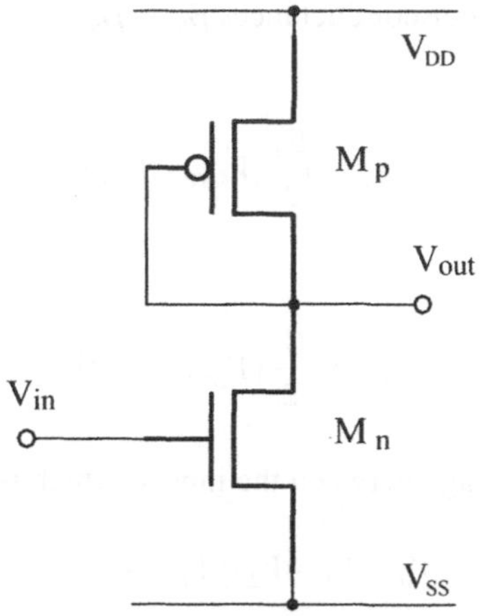

Figure 2.24 Amplifier stage.

The coefficient in the first component of equation (2.81) denotes voltage gain of the stage, whereas the second component denotes voltage DC offset.

2.3.8 Differential Stage

Let us consider the circuit presented in Figure 2.25 and called a differential stage. The transistor *MnB* in this circuit plays the role of a current source that delivers the bias current I_B. Assuming that the transistors *Mn*1 and *Mn*2 are in the saturated

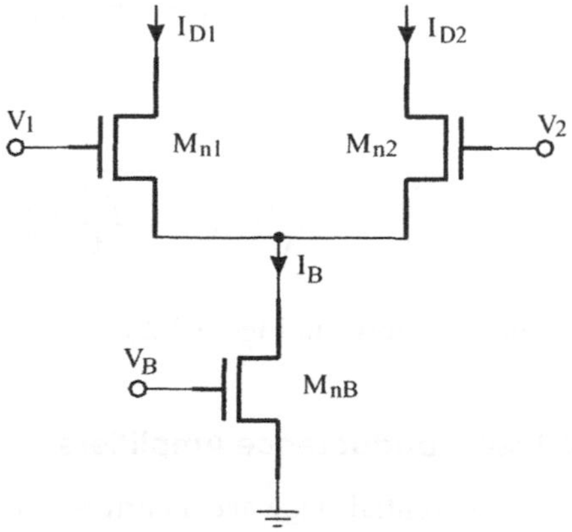

Figure 2.25 Differential stage.

mode and have the same transconductances $\beta_{n1} = \beta_{n2}$, we can express the drain currents as

$$I_{D1} = \frac{\beta_n}{2}(V_{GS1} - V_{Tn})^2 \tag{2.82}$$

and

$$I_{D2} = \frac{\beta_n}{2}(V_{GS2} - V_{Tn})^2 \tag{2.83}$$

Hence, the difference voltage between the inputs, which is defined by

$$V_d = V_1 - V_2 = V_{GS1} - V_{GS2} \tag{2.84}$$

can be written in the form

$$V_d = \sqrt{\frac{2I_{D1}}{\beta_n}} - \sqrt{\frac{2I_{D2}}{\beta_n}} \tag{2.85}$$

or

$$\frac{\beta_n}{2} V_d^2 = I_{D1} + I_{D2} - 2\sqrt{I_{D1} I_{D2}} \tag{2.86}$$

From this equation and from

$$I_{D1} + I_{D2} = I_B \tag{2.87}$$

we can express the drain currents with the use of difference voltage as

$$I_{D1} = \frac{I_B}{2} \pm \frac{1}{2}\sqrt{\beta_n I_B V_d^2 - \frac{\beta_n^2}{4} V_d^4} \tag{2.88}$$

and

$$I_{D2} = \frac{I_B}{2} \mp \frac{1}{2}\sqrt{\beta_n I_B V_d^2 - \frac{\beta_n^2}{4} V_d^4} \tag{2.89}$$

The plots of these currents are shown in Figure 2.26.

2.3.9 Differential Transconductance Amplifiers

A circuit composed of a differential stage and a current mirror with pMOS transistors, presented in Figure 2.27, has the output current obtained as a difference of drain currents (2.88) and (2.89), in the form

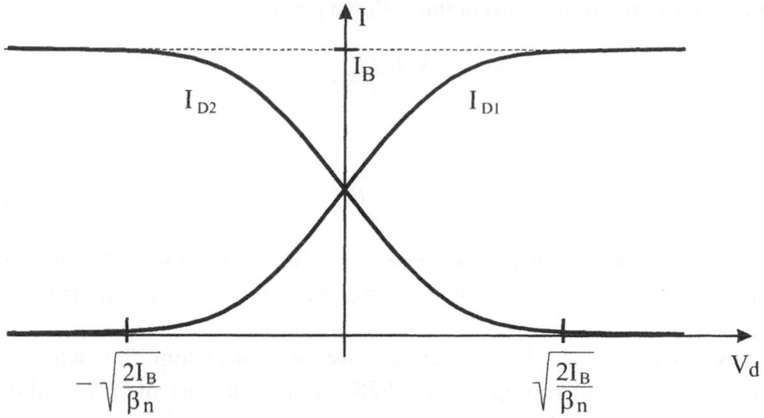

Figure 2.26 Currents in a differential stage.

$$I_{out} = I_{D1} - I_{D2} = \sqrt{\beta_n I_B V_d^2 - \frac{\beta_n^2}{4} V_d^4} \tag{2.90}$$

Assuming that in differential amplifiers

$$\frac{V_d}{2} \ll \sqrt{\frac{I_B}{\beta_n}} \tag{2.91}$$

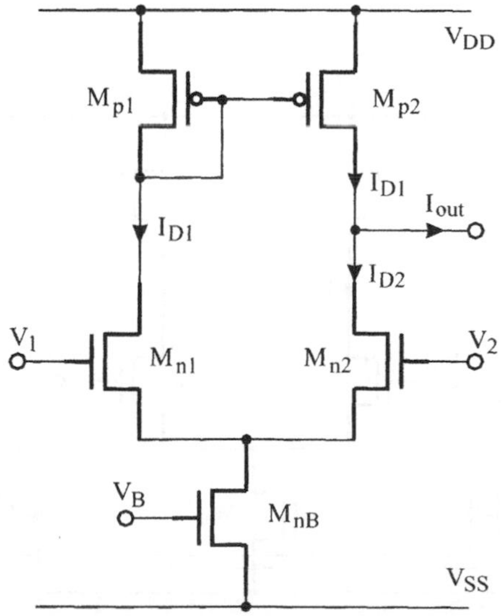

Figure 2.27 Differential transconductance amplifier.

we can now write relation (2.90) in the following way

$$I_{\text{out}} = -\sqrt{\beta_n I_B}(V_2 - V_1) \tag{2.92}$$

where

$$g_m = \sqrt{\beta_n I_B} \tag{2.93}$$

is the differential-mode transconductance of a differential transconductance amplifier. The first input is called the noninverting one, whereas the second is the inverting one.

It is possible to realize fully differential trasconductance amplifiers with two outputs, inverting and noninverting (Figure 2.28). Additional current mirrors allow one to obtain two balanced outputs with the output currents given by relations

$$I_{\text{out+}} = I_1 - I_2 = \sqrt{\beta_n I_B}(V_{\text{in+}} - V_{\text{in-}}) \tag{2.94}$$

and

$$I_{\text{out-}} = I_2 - I_1 = \sqrt{\beta_n I_B}(V_{\text{in-}} - V_{\text{in+}}) \tag{2.95}$$

With the use of fully differential amplifiers, switched capacitor circuits in the balanced structure can be obtained. The advantage of switched current and switched capacitor circuits realized in the balanced structure is the cancellation of parasitic effects.

It is worth mentioning here that the transconductance amplifier described by (2.92) can be used as an analog multiplier. On the basis of (2.5), we can rewrite (2.92) in the form

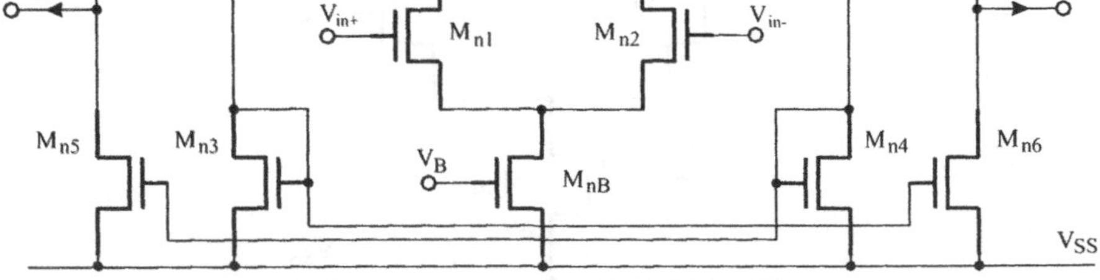

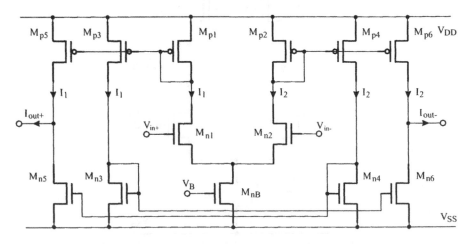

Figure 2.28 Fully differential transconductance amplifier.

$$I_{out} = -\frac{\beta_n}{\sqrt{2}}(V_B - V_{Tn})(V_2 - V_1) \tag{2.96}$$

Denoting

$$V_B = V_{RF}, \qquad V_1 = V_{LO}^+, \qquad V_2 = V_{LO}^- \tag{2.97}$$

we obtain the transconductor in which output current is proportional to the product of the voltages V_{RF} and V_{LO}. Two of such tranconductors that are connected in the balanced structure compose the Gilbert cell [17]. The Gilbert cell is widely used in digital wireless personal communication systems as a so-called mixer. Equations (2.96) and (2.97) show that the mixer is realized as the radio frequency (RF) transconductor in which output current is commutated by the local oscillator (LO).

2.3.10 Operational Amplifiers

The idea of a CMOS operational amplifier (op-amp) is presented in Figure 2.29. The differential stage, composed of pMOS transistors, and a current mirror, composed of nMOS transistors, are components of the first stage of the amplifier. The second stage is composed of a simple voltage amplifier. The compensation feedback branch contains a capacitor and a resistor in series connection. In Figure 2.29 a MOS transistor implements the resistor.

A macro-model of this operational amplifier is shown in Figure 2.30. Node voltage equations of this circuit can be written in the form:

$$\begin{bmatrix} sC_2 + Y + G_2 & -Y + g_{mn3} \\ -Y & sC_1 + Y + G_1 \end{bmatrix} \cdot \begin{bmatrix} V_{out} \\ V_2 \end{bmatrix} = \begin{bmatrix} 0 \\ -g_{mp1} V_{in} \end{bmatrix} \tag{2.98}$$

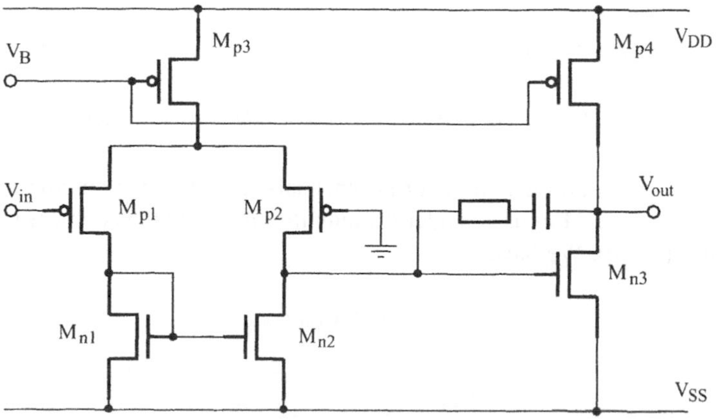

Figure 2.29 CMOS operational amplifier.

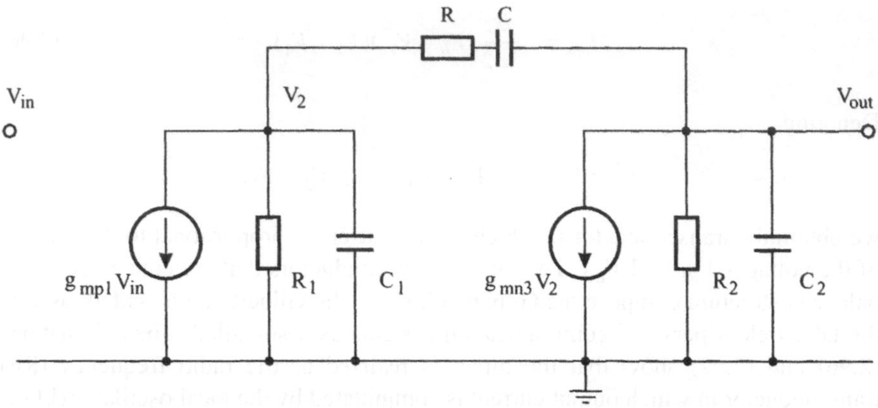

Figure 2.30 Macro-model of a two-stage operational amplifier.

where Y denotes admittance of the compensation branch and has the form

$$Y = \frac{sC \cdot G}{sC + G} \tag{2.99}$$

From the above equations we obtain the transfer function

$$H(s) = \frac{V_{\text{out}}}{V_{\text{in}}} = \frac{N(s)}{D(s)} \tag{2.100}$$

where the numerator polynomial

$$N(s) = g_{mp1}(-Y + g_{mn3}) \tag{2.101}$$

and the denominator polynomial

$$D(s) = s^2 C_1 C_2 + s(C_1 + C_2)Y + s(C_1 G_2 + C_2 G_1) + (G_1 + G_2 + g_{mn3})Y + G_1 G_2 \tag{2.102}$$

The numerator and denominator polynomials show that the op-amp has a single zero and two poles. Let us consider the amplifier in which the compensation branch is omitted ($Y = 0$). We have:

$$N(s) = g_{mp1} g_{mn3} \tag{2.103}$$

and

$$D(s) = s^2 C_1 C_2 + s(C_1 G_2 + C_2 G_1) + G_1 G_2 \tag{2.104}$$

The parasitic capacitance C_1 is much smaller than C_2, in which the load capacitance is included. This implies the inequality

$$C_1G_2 < C_2G_1 \tag{2.105}$$

For inequality (2.105), the transfer function has complex poles, which are zeros of the polynomial (2.104), and the output waveform of the op-amp is oscillating. In order to obtain

$$C_1G_2 = C_2G_1 \tag{2.106}$$

a compensation capacitor with the value approximately equal to the load capacitance can be added in the parallel connection with the capacitance C_1. We have in this case

$$H(s) = \frac{A_o\alpha_p^2}{(s+\alpha_p)^2} \tag{2.107}$$

where

$$\alpha_p = \frac{G_1}{C_1} = \frac{G_2}{C_2} \tag{2.108}$$

is the damping factor and

$$A_o = \frac{g_{mn3}g_{mp1}}{G_1G_2} \tag{2.109}$$

is the dc gain of the op-amp. The unity-gain frequency ω_1 fulfills the equation

$$\frac{A_o\alpha_p^2}{\omega_1^2+\alpha_p^2} = 1 \tag{2.110}$$

giving the gain-bandwidth product, $GBP = \omega_1$, in the form:

$$GBP = \sqrt{A_o - 1}\,\alpha_p \approx A_o\alpha_p \tag{2.111}$$

This compensation method needs a relatively large additional capacitor that significantly increases the chip area.

In order to explain the role of the compensation branch Y, let us assume first that it contains a capacitor, $Y = sC$, exclusively. Hence, the numerator (2.101) and denominator (2.102) polynomials are as follows:

$$N(s) = g_{mp1}(-sC + g_{mn3}) \tag{2.112}$$

and

$$D(s) = s^2(C_1C_2 + C_1C + C_2C) + s[C_1G_2 + C_2G_1 + (G_1 + G_2 + g_{mn3})C] + G_1G_2 \tag{2.113}$$

On the assumptions that $C \gg C_1$, $C_2 \gg C_1$, $g_{mn3} \gg G_1$, and $g_{mn3} \gg G_2$, the denominator polynomial can be approximated as

$$D(s) \approx s^2 C_2 C + s g_{mn3} C + G_1 G_2 \tag{2.114}$$

This polynomial has two real negative zeros, which are poles of the transfer function. Their values can be estimated when the first and third components in (2.114) are omitted, respectively. Hence, the approximate values of poles are as follows:

$$s_p' = -\alpha_p', \qquad s_p'' = -\alpha_p'' \tag{2.115}$$

where

$$\alpha_p' = \frac{G_1 G_2}{C g_{mn3}} = \frac{g_{mp1}}{A_o C}, \qquad \alpha_p'' = \frac{g_{mn3}}{C_2} \tag{2.116}$$

In order to determine *GBP*, the numerator polynomial (2.112) and the denominator polynomial (2.114) with omitted third component will be used for the transfer function $H(j\omega)$ approximation. We then have:

$$H(j\omega) = \frac{g_{mp1}(g_{mn3} - j\omega C)}{j\omega C(j\omega C_2 + g_{mn3})} = \frac{1 - j\omega C/g_{mn3}}{j\omega C/g_{mp1}(1 + j\omega C_2/g_{mn3})} \tag{2.117}$$

and the equation $|H(j\omega_1)| = 1$ has the form:

$$1 + (\omega_1 C/g_{mn3})^2 = (\omega_1 C/g_{mp1})^2[1 + (\omega C_2/g_{mn3})^2] \tag{2.118}$$

Hence, the unity-gain frequency can be estimated by

$$\omega_1^4 \approx \frac{g_{mp1}^2 g_{mn3}^2}{C_2^2 C^2} \tag{2.119}$$

and

$$GBP \approx \sqrt{A_o \alpha_p' \alpha_p''} \tag{2.120}$$

Let us note that in formula (2.116) describing the damping factor α_p', the capacitance C is multiplied by the amplifier DC gain A_o and the compensation can be realized with the use of a small capacitor. This phenomenon is called the Miller effect. A disadvantage of a Miller compensated amplifier is a right-hand plane zero

$$s_z = \frac{g_{mn3}}{C} \tag{2.121}$$

A resistor is typically used in series connection with C. The value of the resistor allows to cancel the zero. It is even used to put the zero into the left-hand plane.

Aside from *GBP*, other important parameters that describe the op-amp behavior are: phase margin (*PM*), slew rate (*SR*), settling time (t_s), power supply rejection ratio (*PSRR*), and common mode rejection ratio (*CMRR*). These parameters can be expressed in the following way:

1. $PM = 180° - \arg[H(j\omega_1)]$, where ω_1 is the unity-gain frequency
2. $SR = dV_{out}/dt$, which is the maximum rate of V_{out} change
3. $t_s = (3 \div 5)\tau$, where the time constant $\tau = 1/\alpha$ is calculated for being the smallest damping factor α
4. $PSRR = 20 \log (|H_o(j\omega)|/|H_{supp}(j\omega)|)$, where $H_o(j\omega)$ is an open loop transfer function of an amplifier and $H_{supp}(j\omega)$ is the supply-to-output transfer function
5. $CMRR = 20 \log (|H_o(j\omega)|/|H_c(j\omega)|)$, where $H_o(j\omega)$ is an open loop transfer function of an amplifier and $H_c(j\omega)$ is a common-mode transfer function which can be measured for excitation delivered to both inverting and noninverting inputs simultaneously

Denoting by Q_{ch} the charge delivered to the capacitor C, the above definition of slew-rate can be replaced, on the basis of the relation $Q_{ch} = CV_{out}$, by $SR = I_{ch}/C$, where I_{ch} is the amount of current needed to charge the compensation capacitor C. The settling time t_s, defined as a multiple of the time constant τ, is applicable in linear circuits in which the decrease of voltages is described by the exponential function $v(t) = V_0 \exp(-t/\tau)$. In nonlinear circuits in which voltages can described by other functions, the settling time t_s is defined as a time interval in which the voltage decreases to the level specified as a percentage of its initial value.

2.4 PROBLEMS

1. For the switch presented in Figure 2.7a, terminated with resistor R, and for the data $V_{DD} = 1.5$ V, $V_{SS} = -1.5$ V, $k'_n = 5 \cdot 10^{-5}$A/V^2, $V_{Tn} = 1$ V, $W = 10$ μm, $L = 1$ μm, $R = 1$ kΩ, $V_{Tn} = 1$ V, calculate the maximum value of the output voltage V_{out}. Consider two cases of the pulse applied to the input of the switch:
   (a) the positive pulse with the maximum value V_{DD}
   (b) the negative pulse with the minimum value V_{SS}
2. Calculate the integral (2.36) of the differential equation (2.32) with the use of substitution (2.34).
3. Assume that V_{in} in Figure 2.9 does not reach a sufficiently big value to introduce the transistor into the saturated mode. Estimate the time constant τ_{ch} in this case. Is the inequality (2.38) still valid under this assumption?
4. Prove that the saturation of a pMOS transistor operating in the inverter occurs at the point given by the formula $V_{Ip} = V_{Tp}/(1 + \mu)$.
5. Using equations (2.59) and (2.60) and (2.61) and (2.62), calculate V_{IL}, V_{IH},

V_{OL}, V_{OH}, and V_I, V_{TW}, V_{NML}, V_{NMH} for $V_{DD} = 3$ V, $k_n' = 5 \cdot 10^{-5}$ A/V^2, $k_p' = 2.5 \cdot 10^{-5}$A/V^2, $V_{Tn} = 1$ V, $V_{Tp} = -0.8$V and $W = 10$ μm, $L = 1$ μm for both kinds of transistors.

6. On the basis of relations (2.70) and (2.71), calculate the widths of nMOS and pMOS transistors assuming that their length is $L = 2$ μm and $V_{DD} = 3$ V, $k_n' = 5 \cdot 10^{-5}$A/V^2, $k_p' = 2.5 \cdot 10^{-5}$A/V^2, $V_{Tn} = 1$ V, $V_{Tp} = -0.8$ V, $\tau_n = 2$ ns, $\tau_p = 3.64$ ns, $C_{out} = 1$ pF. Determine the maximum switching frequency f_{max}.
7. Find the relations (2.88) and (2.89) describing the drain currents of a differential stage as the solution of equations (2.86) and (2.87).
8. Calculate the poles and the zero of the transfer function (2.100) for $R = 0$ and the zero for $R \neq 0$.

3

CMOS Circuit Fabrication

The basic processes of integrated circuit fabrication in a foundry are

- Wafer preparation
- Oxidation
- Deposition
- Lithography and etching
- Epitaxy, diffusion, and ion implantation

The designer of integrated circuits ought to be familiar with these processes in order to understand the project steps and the design rules and to be able to modify the operations that violate these rules. This chapter contains a brief description of the basic processes listed above. The reader can find more about this technology in [12, 16, 60].

3.1 WAFER PREPARATION

CMOS technology uses single-crystal silicon wafers for integrated circuit fabrication. Wafers are silicon disks 75 mm to 230 mm in diameter. Silicon in its polycrystal form is obtainable from sand. In the wafer production process, the level of natural impurities is lowered and other substances are added, in strictly controlled amounts, to obtain the single-crystal form with the required electrical properties. Wafers are less than 1 mm thick and are sliced from cylindric ingots pulled from silicon melt in the process known as the Czochralski method. One face of the wafer, on which the integrated circuit will be built, is polished before further operation.

Professor Jan Czochralski, a Polish scientist, published his method in the German journal "Zeitschrift für physikalische Chemie" in 1918. However, he was known mainly for "Bahnmetal" (the metal for train bearings), which was strategic material in Germany during the First World War. This invention made him a rich man. He came back to Poland after the rebirth of the state. In Poland, he was able to very effectively combine his scientific work with activity in industry. His excellent career ended with the Second World War.

3.2 OXIDATION

Silicon dioxide (SiO_2) is the natural insulator material of the silicon-based semiconductor industry. As can be seen in Figure 2.1, which shows the structure of a transistor, during oxidation it is important to obtain a very thin film of so-called gate oxide where the gate will be laid and the transistor will be located. Thin layers of SiO_2 are covered by patterned photoresist and another layer of dioxide is grown. The patterned regions are covered by gate oxide and are called active areas, whereas the remaining area is covered by field oxide.

3.3 DEPOSITION, LITHOGRAPHY, AND ETCHING

Lithography and etching are processes similar to those known from photographic techniques. They are used in order to obtain necessary patterns in films. The film materials are usually silicon dioxide (SiO_2), silicon nitride (SiN), polysilicon (polycrystalline silicon), photoresist (photoresistive organic material) and metal (aluminum and occasionally copper). These materials are deposited on the wafer. Chemical vapor deposition (CVD) is a frequently used deposition technique.

Optical photolithography, which uses ultraviolet light, is the common technique used to obtain patterns in a photoresist. The layer deposited on the wafer and not protected by the photoresist is etched by appropriate chemicals. For higher resolution of the lithography process, an electron beam or an X-ray source can be used instead of an ultraviolet light source.

3.4 EPITAXY, DIFFUSION, AND ION IMPLANTATION

Silicon in the single-crystal form is an insulator. Two kinds of impurities—acceptors and donors—are introduced into this material in order to obtain a conducting *p*-type and *n*-type material, respectively. Epitaxy, diffusion, and ion implantation processes are used for the introduction of impurities.

The epitaxial layer is obtained during the epitaxial growth. The CVD technique is used in the process. Epitaxial layers are usually used in bipolar transistor fabrication.

Diffusion is the method most frequently used for doping silicon in chosen areas of the wafer. Impurities diffuse from the layer deposited on the surface into the silicon and inside the silicon between areas with different concentrations of dopants. This kind of migration also takes place in integrated circuits at normal operating temperatures. However, it takes tens of years for natural migration to become significant.

In the ion implantation method, ionized impurities are accelerated and lodged in the substrate. Dopant concentration (in the substrate) depends on the velocity and density of the ionized impurity beam. This method ensures good control of dopant concentration.

3.5 CONTACTS AND INTERCONNECTS

Interconnects between elements of a circuit are etched in polysilicon and metal films. Polysilicon is used for gate electrodes of transistors and short interconnects. Connections through isolating layers are called contacts (connections between metal and doped silicon) or vias (connections between metal layers).

3.6 MASKS AND DESIGN RULES

Photolithography is used in manufacturing integrated circuits. Hence, the description of mask geometries for all lithography processes ought to be the result of integrated circuit design. It is easy to obtain such description using current computer systems called silicon compilers. Masks in a given format (usually the Caltech Intermediate Format or CIF is used) are sent to a foundry where a laser beam or an electron beam are used to produce actual masks.

In order to illustrate the use of masks in the fabrication of CMOS circuits, let us describe an n-well, single-poly, double-metal process. The description of the process is simplified and some intermediate steps are omitted in order to emphasize the role of each mask. The process is as follows:

00 start with p-type wafer
01 diffuse n-well, mask: *01*, **nwell**
02 grow gate oxide
03 define active area, mask: *02*, **active**
04 grow field oxide
05 deposit polysilicon
06 pattern polysilicon, mask: *03*, **poly**
07 implant n^+ dopants, mask: *04*, **nplus**
08 implant p^+ dopants, mask: *05*, **pplus**

09 pattern contact openings, mask: *06*, **contact**
10 deposit metal 1
11 pattern metal 1, mask: *07*, **metal1**
12 deposit oxide (CVD)
13 pattern metal 2 contacts, mask: *08*, **via**
14 deposit metal 2
15 pattern metal 2, mask: *09*, **metal2**
16 deposit glass (nitride passivation)
17 pattern pad openings, mask: *10*, **pad**

Minimum sizes of the patterns on the masks depend on the resolution of the lithography and on the features of the technology process. The set of geometrical specifications of patterns is called the design rules. The process-specific design rule set contains all dimensions in microns and can be applied to a particular fabrication process only. A more general approach is scaling, in which a generic metric λ is used. All dimensions are given in multiples of λ, which makes the layout independent of the fabrication process. When the process is chosen, λ assumes a value; for example 0.2 μm in the 0.4 CMOS process.

Basic design rules are shown in Figure 3.1. The minimum widths, spacings and other dimensions shown in Figure 3.1 (in multiples of λ and in microns for a hypothetical 0.4 CMOS *p*-substrate (*n*-well) process) are as follows:

w1 *n*-well width, $w1 = 6\lambda = 1.2\ \mu m$
w2 *n*-well spacing, $w2 = 5\lambda = 1\ \mu m$
w3 *n*-well spacing to *n*-plus (*p*-plus), $w3 = 6\lambda = 1.2\ \mu m$
d1 *n*-plus (*p*-plus) width, $d1 = 3\lambda = 0.6\ \mu m$
d2 *n*-plus (*p*-plus) spacing, $d2 = 3\lambda = 0.6\ \mu m$
p1 polysilicon width, $p1 = 2\lambda = 0.4\ \mu m$
p2 polysilicon spacing, $p2 = 2\lambda = 0.4\ \mu m$
p3 polysilicon spacing to *n*-plus (*p*-plus), $p3 = 1\lambda = 0.2\ \mu m$
p4 polysilicon overhang (extension of gate), $p4 = 2\lambda = 0.4\ \mu m$,
c1 contact dimension, $c1 = 2\lambda = 0.4\ \mu m$
c2 contact spacing, $c2 = 2\lambda = 0.4\ \mu m$
c3 metal enclosure of contact, $c3 = 1\lambda = 0.2\ \mu m$,
m1 metal width, $m1 = 3\lambda = 0.6\ \mu m$
m2 metal spacing, $m2 = 4\lambda = 0.8\ \mu m$

Complete design rules can be obtained from the foundries in technological files, in a format suitable for tools used by designers of integrated circuit layouts.

Figure 3.2 shows the layout of the inverter from Figure 2.15, illustrating the

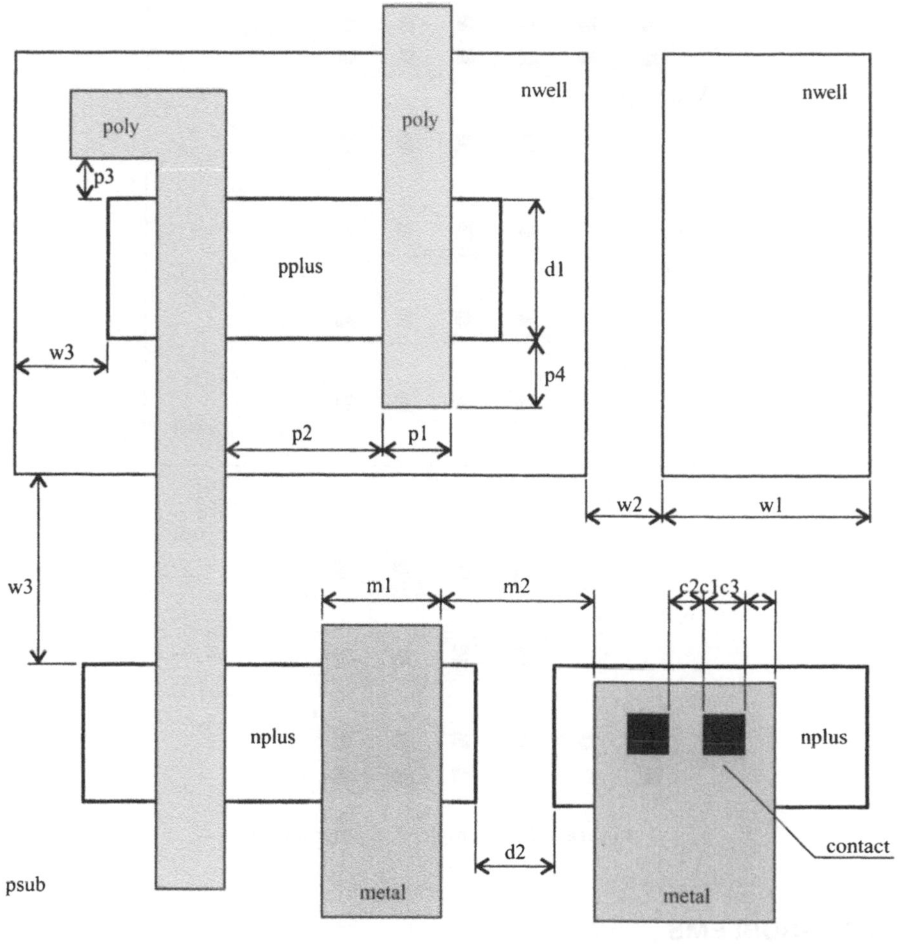

Figure 3.1 Example of design rules.

above design rules. The lengths of transistor channels are many times greater than their widths. Hence, the transistors are laid out as folded devices. The use of folded transistors is characteristic of analog cells in which such channel dimensions are typical. In Figure 3.2, the transistor *Mp* is folded three times and *Mn* is folded once. In comparison to unfolded transistors, folded devices have smaller areas of drains and sources, which decreases parasitic effects. Polysilicon, depicted by the shaded area, forms the gates of the transistors and is used as the connection between the gates and the input. The areas of the sources and drains are not visible in Figure 3.2. Their location is pointed by black boxes, which are contacts with *p*-plus and *n*-plus regions. The metal V_{DD} line has additional contacts with the *n*-well area.

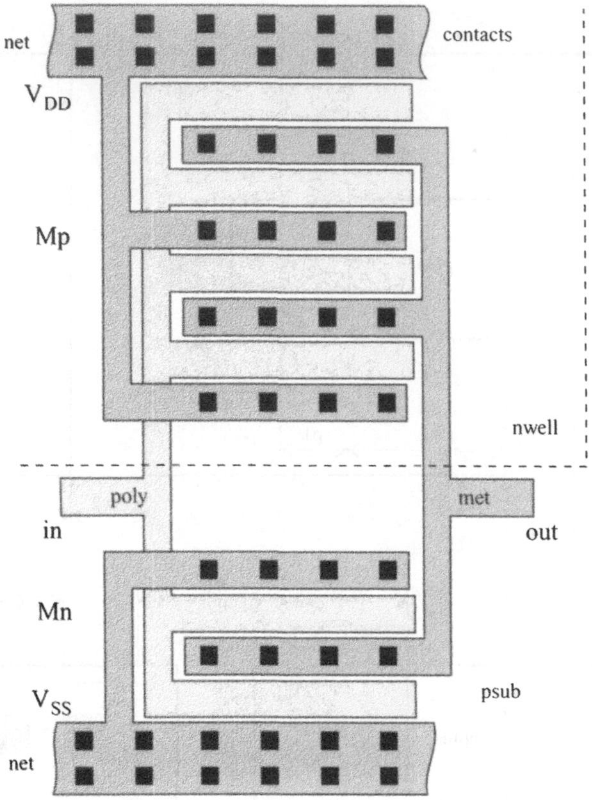

Figure 3.2 Example of inverter layout.

3.7 PROBLEMS

1. Using the design rules for a hypothetical 0.4 CMOS *p*-substrate (*n*-well) process, draw a layout of the switch in Figure 2.10, assuming that the widths and lengths of transistor channels are as small as possible.
2. Draw a layout of the Schmitt trigger shown in Figure 2.19, for a hypothetical 0.4 CMOS *p*-substrate (*n*-well) process. Assume that widths of pMOS transistor channels are 24 μm and those for nMOS are 8 μm, whereas all channel lengths are 4 μm.

4

Digital Techniques

Digital techniques are described in many excellent books. In this chapter we will describe the digital techniques that are very frequently used in mixed signal integrated circuit design. In particular, we will describe static and dynamic logic gates and, briefly, finite-state machines and memories.

4.1 STATIC LOGIC CIRCUITS

The general structure of a static logic circuit is presented in Figure 4.1. The inverter shown in Figure 2.15 is a special case of a static logic circuit that realizes the logic function $F = x_1$. The blocks of logic functions F and F act as mutually exclusive switches, as the nMOS and pMOS transistors in the inverter. In other words, the nMOS and pMOS logic arrays play the role of respective single transistors in the inverter (Figure 2.15). Depending on the logic values of the input variables $x_1, \ldots,$ x_n, the logic function F can achieve a logic values 0 or 1. If $F = 1$ $(\overline{F} = 0)$ then the nMOS array is conducting, whereas the pMOS array acts as an open circuit. In this case, the output voltage is low and the logic response $y = 0$. If $F = 0$ $(\overline{F} = 1)$, then the nMOS array is an open circuit and the pMOS array is a short circuit, resulting in high output voltage and logic response $y = 1$.

4.1.1 NAND and NOR Gates

The inverter is the only logic circuit in which pMOS and nMOS logic arrays are composed of a single transistor each. Logic circuits containing arrays composed of two transistors in a series or parallel connection are shown in Figure 4.2a,b. In the

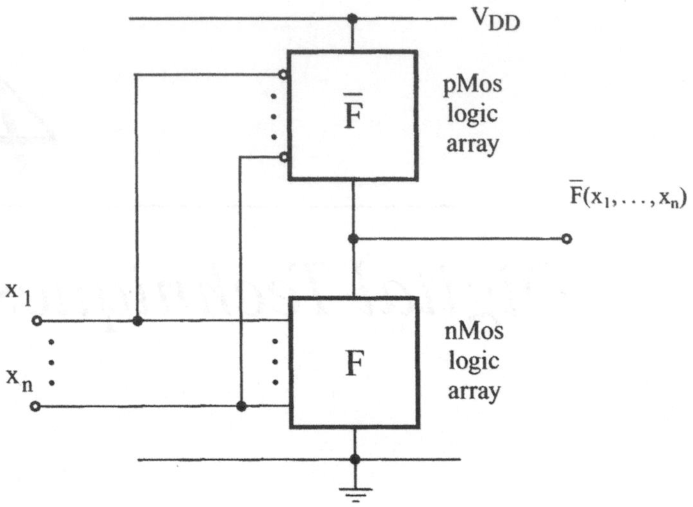

Figure 4.1 General structure of a static logic circuit.

first logic circuit, the nMOS logic array is conducting when both input signals x_1 and x_2 are equal to 1. In the second circuit, the nMOS array is conducting when at least one of input signals is equal to 1. Hence, the first array realizes the *and* (x_1x_2) operation and the second the *or* $(x_1 + x_2)$ operation. As a result, we obtain *NAND* and *NOR* gates with the output signals

$$y = \overline{x_1 x_2}, \qquad y = \overline{x_1 + x_2} \tag{4.1}$$

In order to estimate transient characteristics of NAND and NOR gates we can use the formulae (2.67) and (2.69) for the transistor channel resistances of nMOS and pMOS transistors, respectively. However, logic arrays are now composed of two transistors in a parallel or series connection. Hence, resistances in relations (2.66) and (2.67) must be replaced by the resultant resistances of two transistors. Consistently, the formulae (2.70) and (2.71) should be modified depending on the kind of the connection.

As an example, let us consider the charging process in the NAND gate. Let us assume that both pMOS transistors are identical. The worst-case situation occurs when only one transistor is conducting. In this case we have

$$\left(\frac{W}{L}\right)_p = \frac{C}{k_p' \tau_p (V_{DD} + V_{Tp})} \tag{4.2}$$

where $\tau_p \approx t_{LH}/3$ and t_{LH} is the low-to-high time. Similarly, for the high-to-low response we consider the discharging processes through the series-connected nMOS transistors and we obtain

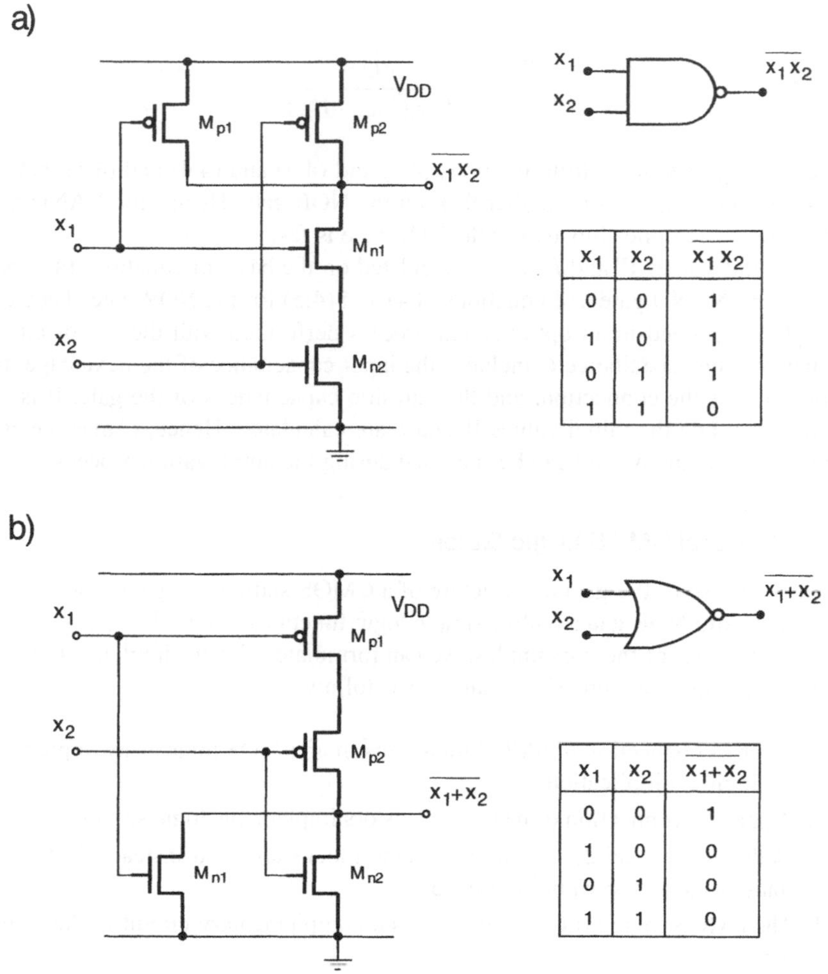

x_1	x_2	$\overline{x_1 x_2}$
0	0	1
1	0	1
0	1	1
1	1	0

x_1	x_2	$\overline{x_1 + x_2}$
0	0	1
1	0	0
0	1	0
1	1	0

Figure 4.2 NAND (a) and NOR (b) gates with their symbols and truth tables.

$$\left(\frac{W}{L}\right)_n = \frac{2C}{k'_n \tau_n (V_{DD} - V_{Tn})} \tag{4.3}$$

where $\tau_n \approx t_{HL}/3$ and the factor 2 in the numerator arises as the result of the above-mentioned series connection of the transistors.

For the NOR gate we can obtain the complementary relations

$$\left(\frac{W}{L}\right)_p = \frac{2C}{k'_p \tau_p (V_{DD} + V_{Tp})} \tag{4.4}$$

and

$$\left(\frac{W}{L}\right)_n = \frac{C}{k'_n \tau_n (V_{DD} - V_{Tn})} \tag{4.5}$$

Since $k'_n > k'_p$, it follows from relations (4.2) and (4.3) and (4.4) and (4.5) that for the NAND gate $t_{LH} + t_{HL}$ is smaller than for the NOR gate. Hence, the NAND gate that occupies the same chip area as the NOR gate is faster.

The initial values W and L can be calculated on the basis of equations (4.2) and (4.3) for the NAND gate and equations (4.4) and (4.5) for the NOR gate. They can be further improved in an optimization process performed with the use of circuit simulators. The capacitance C includes the input capacitance of the next stage, the capacitance of the connection, and the parasitic capacitances of the gate. It is not determined when the initial values W and L are calculated. Hence, C must be estimated at the beginning and can be specified during the optimization process.

4.1.2 General CMOS Logic Gates

Figure 4.1 presents the general structure of a CMOS static logic gate. The inverter and NAND and NOR gates realize simple logic functions within this general structure. On the basis of these examples, we can formulate rules for implementation of an arbitrary logic function. These rules are as follows:

1. A series connection of nMOS transistors in an nMOS logic array implements the logic multiplication.
2. A parallel connection of nMOS transistors implements logic summation.
3. Both rules, 1 and 2, also apply if logic blocks are used instead of of single transistors, as shown in Figure 4.3.
4. The pMOS logic array is obtained as a complementary circuit to the nMOS array.
5. The output of the logic element designed is the complement of the function realized by the nMOS logic array.

We shall illustrate these rules by implementing an exclusive-OR (XOR) function, described by the logic function in the form:

$$F = x_1 \oplus x_2 = x_1\overline{x}_2 + \overline{x}_1 x_2 \tag{4.6}$$

The symbol and the truth table of an XOR element are shown in Figure 4.4. We see from the truth table that the output of XOR is "0" when $x_1 = x_2$ and "1" when either $x_1 = 1$ or $x_2 = 1$ exclusively. In order to realize the nMOS logic array of XOR, we will use the complement of the function (4.6):

$$G = \overline{F} = (\overline{x}_1 + x_2)(x_1 + \overline{x}_2) = \overline{x}_1\overline{x}_2 + x_1 x_2 \tag{4.7}$$

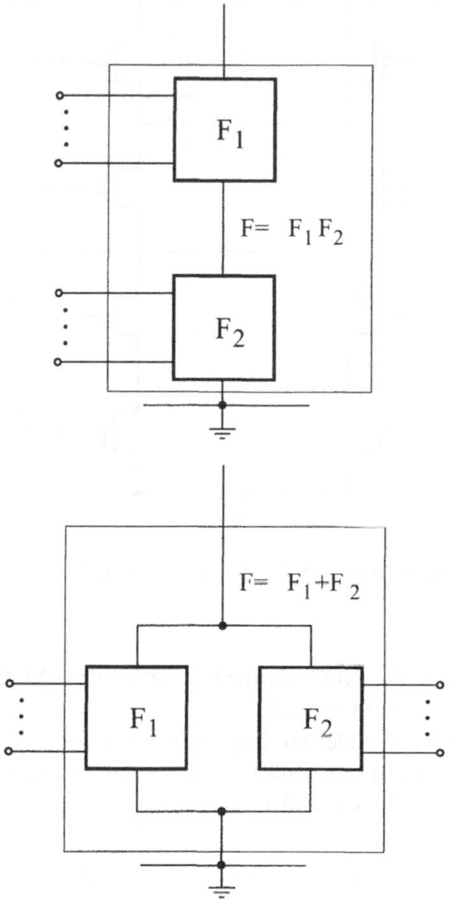

Figure 4.3 Implementation of the logic functions $F = F_1F_2$ and $F = F_1 + F_2$ by the series and parallel connection of logic blocks.

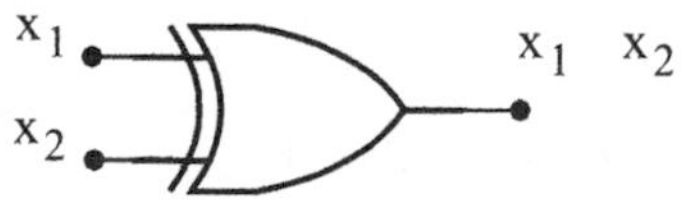

x_1	x_2	$x_1 \; x_2$
0	0	0
1	0	1
0	1	1
1	1	0

Figure 4.4 Symbol and truth table of an XOR element.

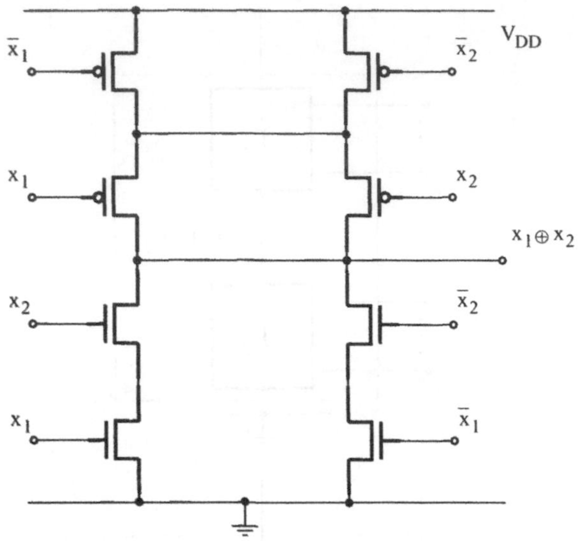

Figure 4.5 Implementation of an XOR element.

The implementation of the XOR element in which the nMOS logic array realizes the function G is shown in Figure 4.5.

A half-adder and a full-adder are important applications of the XOR element. A half-adder calculates the sum s_0 and the carry c_0 of less significant bits (LSB) x_{10} and x_{20} of two n-bit numbers x_1 and x_2 as

$$s_0 = x_{10} \oplus x_{20}, \qquad c_0 = x_{10}x_{20} \tag{4.8}$$

Sum and carry bits of remaining bits $x_{1j}, x_{2j}, j = 1, \ldots, n-1$ are calculated as

$$s_j = x_{1j} \oplus x_{2j} \oplus c_{j-1}, \qquad c_j = x_{1j}x_{2j} + x_{1j}c_{j-1} + x_{2j}c_{j-1}, \qquad j = 1, \ldots, n-1 \tag{4.9}$$

One of possible implementations of the full-adder, shown in Figure 4.6, is obtained after transformation of the logic functions s_j and c_j in (4.9) into the form

$$\begin{aligned} s_j &= (x_{1j} + x_{2j} + c_{j-1})\bar{c}_j + x_{1j}x_{2j}c_{j-1} \\ c_j &= x_{1j}x_{2j} + c_{j-1}(x_{1j} + x_{2j}), \qquad j = 1, \ldots, n-1 \end{aligned} \tag{4.10}$$

4.1.3 Pseudo-nMOS and Pseudo-pMOS Logic

Static CMOS logic gates can be replaced with pseudo-nMOS or pseudo-pMOS devices in order to reduce the number of MOS transistors and, consequently, the chip

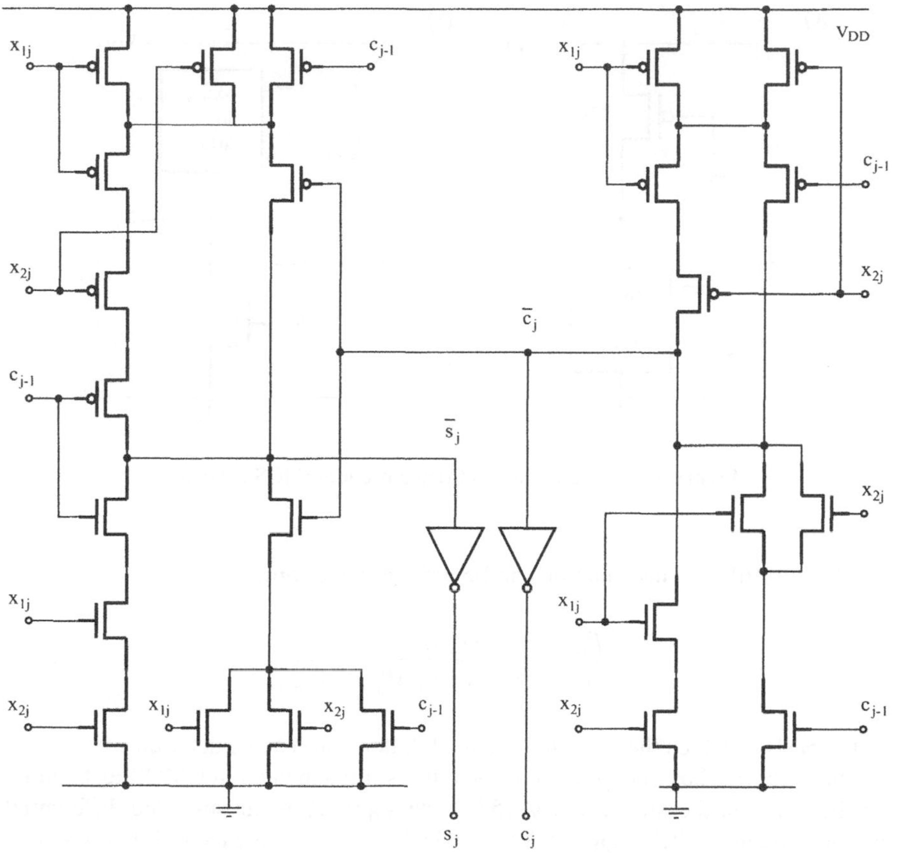

Figure 4.6 Implementation of a full-adder.

area and power consumption. Logic gates considered in this section achieve this goal, as they consist of only one logic array, either nMOS or pMOS, as shown in Figure 4.7a,b. The second array that appears in CMOS logic gates is now replaced by a single pMOS or nMOS transistor which is the load of the driver array. Let us explain this idea using as an example the pseudo-nMOS inverter presented in Figure 4.8. The pMOS transistor is biased on. When the logic value "0" is applied to the input, then the nMOS transistor is off and the voltage on the output $V_{OH} = V_{DD}$, which is associated with the logic value "1." For the input corresponding to "1," both transistors are on and the pMOS transistor is saturated, whereas the nMOS one is nonsaturated. Hence, for the desired low voltage V_{OL} at the output the following equation for the drain currents

$$\frac{\beta_n}{2}[2(V_{DD} - V_{Tn})V_{OL} - V_{OL}^2] = \frac{\beta_p}{2}(V_{DD} + V_{Tp})^2 \tag{4.11}$$

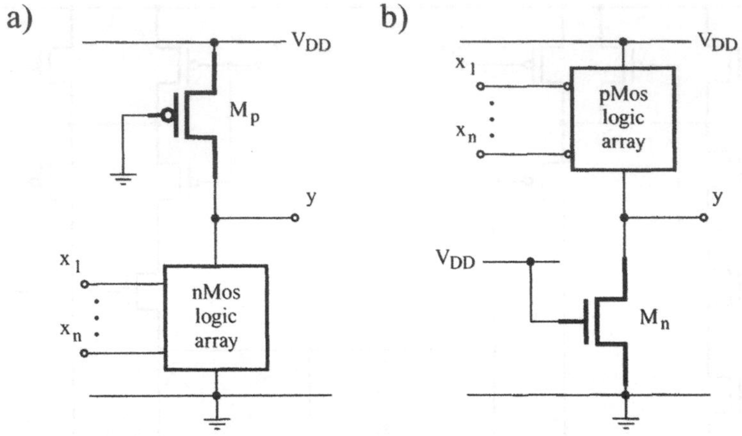

Figure 4.7 Pseudo-nMOS (a) and pseudo-pMOS (b) gates.

should be fulfilled. This equation can be written in the form

$$\frac{\beta_n}{\beta_p} = \frac{(V_{DD} + V_{Tp})^2}{2(V_{DD} - V_{Tn})V_{OL} - V_{OL}^2} \tag{4.12}$$

on the basis of which the transistor channel dimensions can be determined.

Other approaches to digital circuit synthesis implemented in CMOS technology can be found in the literature, e.g. [57]. One approach is the so-called differential cascode voltage switch logic, DCVS, in which two cross-coupled pMOS transistors

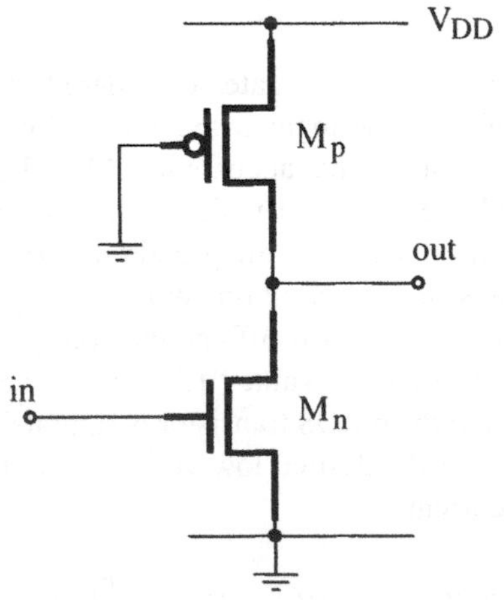

Figure 4.8 Pseudo-nMOS inverter.

are introduced as a load. Differential split-level (DSL) can be obtained as the modified DCVS logic.

4.1.4 Flip-Flops

It is possible to construct all types of flip-flops on the basic of the static logic gates introduced above. However, flip-flops based on transmission gates (TG), for example those shown in Figure 2.21, have better performance. Hence, in this section we will consider only two flip-flops based on static logic circuits (two SR flip-flops shown in Figure 4.9 and Figure 4.10). The remaining flip-flops considered here will contain transmission gates.

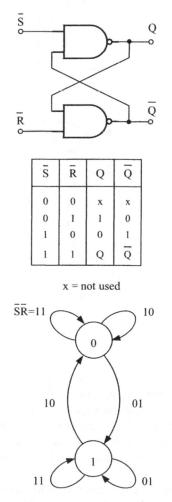

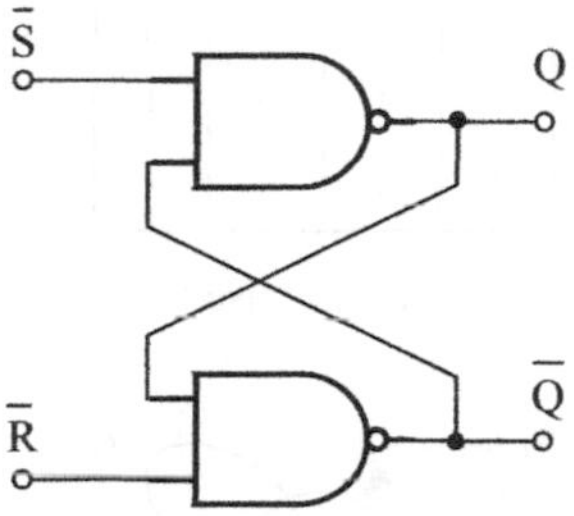

$\bar{S}$	$\bar{R}$	Q	$\bar{Q}$
0	0	x	x
0	1	1	0
1	0	0	1
1	1	Q	$\bar{Q}$

x = not used

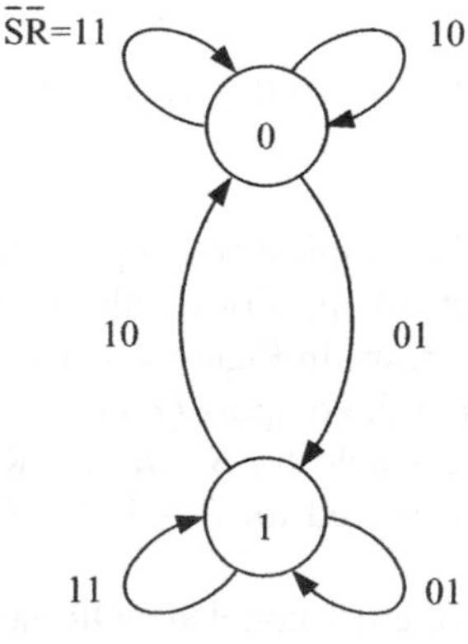

Figure 4.9 SR flip-flop based on NAND gate.

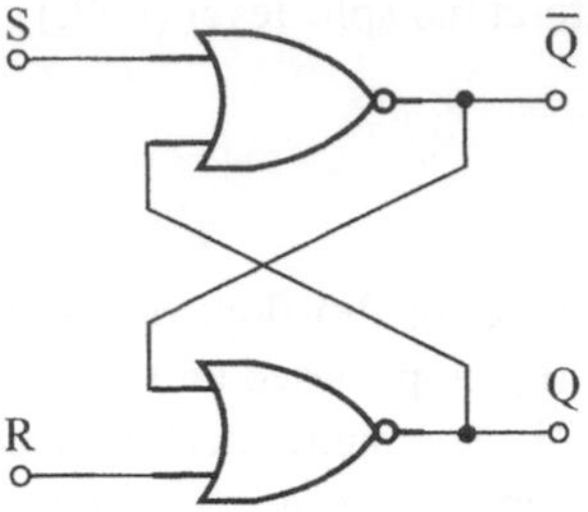

S	R	Q	$\bar{Q}$
0	0	Q	$\bar{Q}$
1	0	1	0
0	1	0	1
1	1	x	x

x = not used

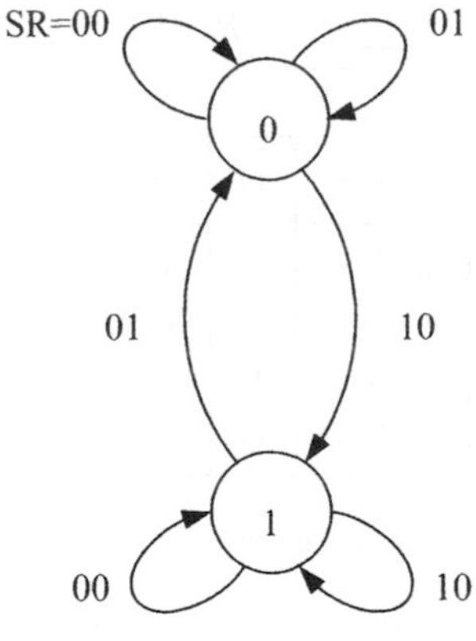

Figure 4.10 SR flip-flop based on NOR gate.

4.1.4.1 SR Flip-Flops The simplest two-input static logic gates, NAND and NOR, can be used to construct SR flip-flops with the inputs S (set) and R (reset). The NAND-based flip-flop is shown in Figure 4.9. The input state $S = R = 1$ is not used because of contradiction at the outputs Q and $\bar{Q}$, which are not complementary for $S = R = 1$. The output is held for $S = R = 0$. Resetting $Q = 0$, and setting $Q = 1$, of the SR flip-flop is achieved for $R = 1$, $S = 0$ and $S = 1$, $R = 0$, respectively.

An SR flip-flop composed of cross-coupled NOR gates is shown in Figure 4.10. In this case, the input state $S = 1$, $R = 1$ is not used. The output state is held for $S = 0$, $R = 0$.

Table 4.1 Operation of a toggle flip-flop

		Q'	
Q	$\bar{S}$	$\Phi = 0$	$\Phi = 1$
0	0	1	1
0	1	0	1
1	0	1	1
1	1	1	0

4.1.4.2 Toggle and JK Flip-Flop The DFF shown in Figure 2.21 can be developed with the use of NAND or NOR gates, as it is shown in Figure 4.11 and Figure 4.12. The first flip-flop, in which two inverters are replaced by NAND gates and the feedback branch is added, is called a toggle flip-flop (TFF). Let us analyze the operation of the TFF using Table 4.1. The first column in Table 4.1 shows the current state of the TFF. The second column shows the set signal. The last column shows the next state of the TFF in the first ($\Phi = 0$) and second ($\Phi = 1$) half of the clock period. On the basis of Table 4.1, the state diagram shown in Figure 4.11 can be obtained.

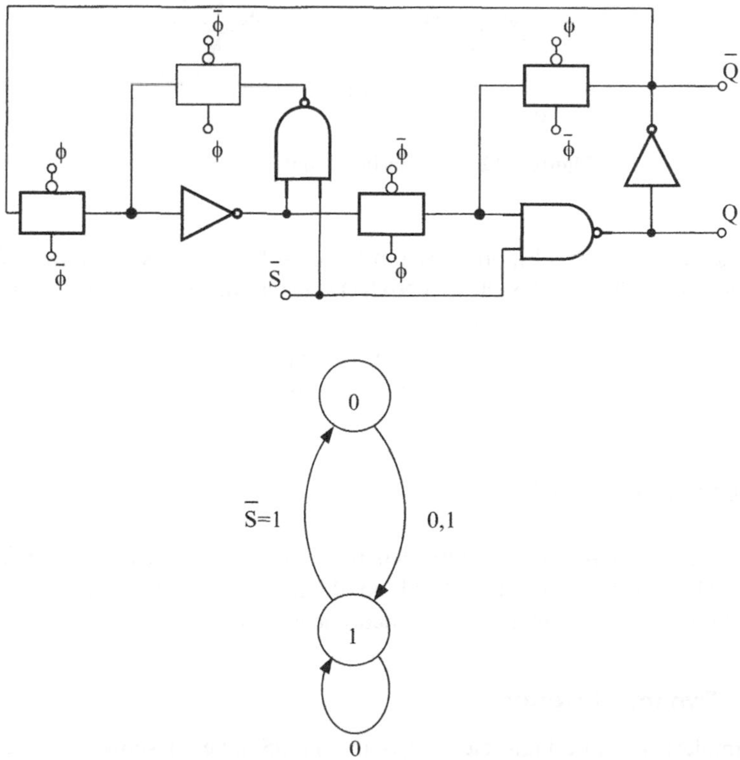

Figure 4.11 Toggle flip-flop and its state diagram.

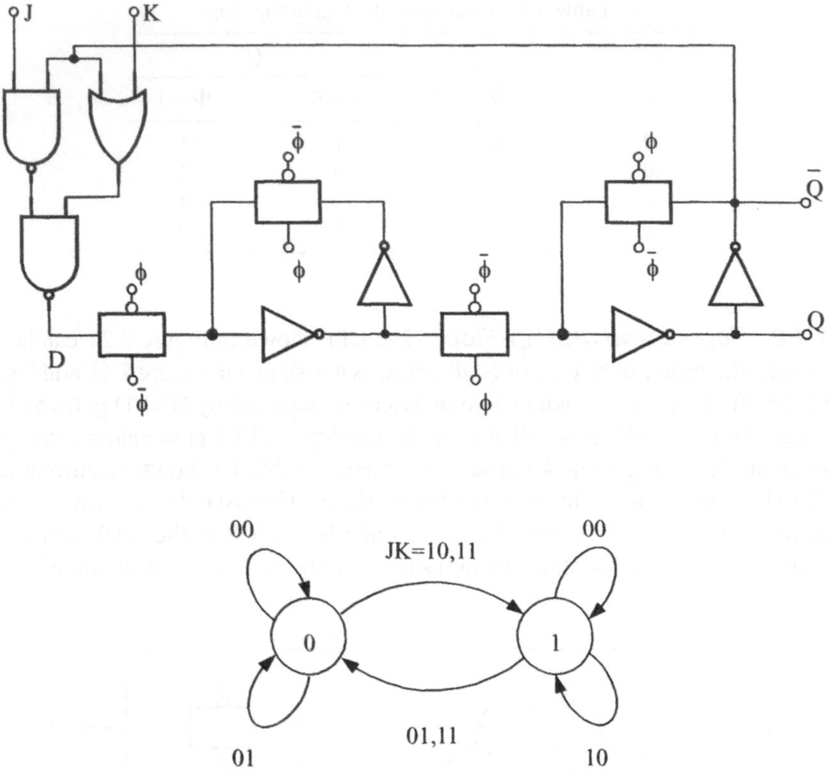

Figure 4.12 JK flip-flop and its state diagram.

The JK flip-flop (JKFF), presented in Figure 4.12, contains two inputs, J and K, introduced with the use of NOR and NAND gates in the feedback branch. The combinational part of the flip-flop, composed of NOR and NAND gates, is described by the function $D = \overline{J\overline{Q}(K+\overline{Q})} = J\overline{Q} + \overline{K}Q$. The state diagram shown in Figure 4.12 is obtained as a result of the assumption that the next state is equal to the logic value D.

4.2 DYNAMIC LOGIC

Dynamic digital circuits are controlled by a two-phase clock, like the flip-flops composed of transmission gates. The kind of logic in which the operation of devices is synchronized by a clock is called synchronous logic.

4.2.1 Dynamic Inverter

The simplest dynamic logic element is the nMOS inverter shown in Figure 4.13. The clock separates the operation of the inverter into two phases: the precharge in-

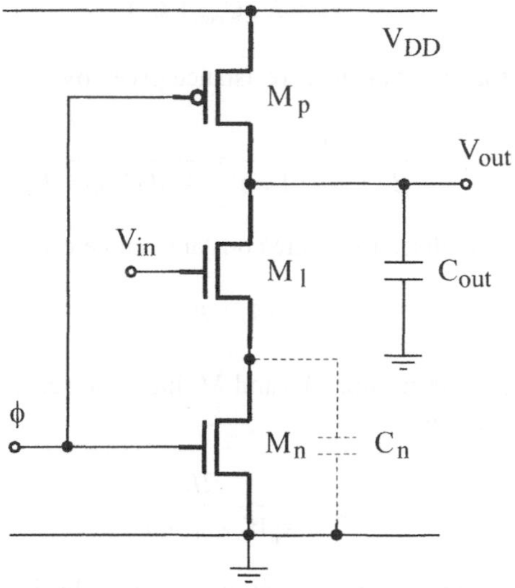

Figure 4.13 Dynamic nMOS inverter.

terval and the evaluate interval. The parasitic capacitors C_{out} and C_n of the corresponding nodes in the circuit in Figure 4.13 are charged or discharged in the precharge and evaluate intervals. The phase $\Phi = 0$, in which the transistor M_p is on and the M_n one is off, corresponds to the precharge interval. In this phase the capacitor C_{out} is charged to the voltage level $V_{out} = V_{DD}$, regardless of the input signal V_{in}. In the second phase, $\Phi = 1$, the transistor M_p is off and the M_n one is on, and the operation of the circuit depends on the status of the logic transistor M_l. When the input voltage is low, $V_{in} = V_{IL}$, corresponding to a logic "0," the M_l transistor is off, and the output signal is still a logic "1," achieved in the precharge phase. However, for the input signal corresponding to a logic "1," both transistors M_n and M_l conduct and the capacitor C_{out} is discharged. The output is a logic "0." The maximum clock frequency f_{max} is determined by t_{max}

$$f_{max} = \frac{1}{2t_{max}} \tag{4.13}$$

where

$$t_{max} = (3 \div 5)\max(\tau_{ch}, \tau_{dis}) \tag{4.14}$$

and τ_{ch}, τ_{dis} are charge and discharge time constants. The time constants depend on the design parameters, the width W, and the length L of transistor channels.

The worst-case situation in the precharge interval occurs when both capacitors C_{out} and C_n are charged. In order to obtain a simple formula, the resistance R_l of the M_l transistor channel is neglected, and the charge time constant τ_{ch} is described by

$$\tau_{\text{ch}} = R_p(C_{\text{out}} + C_n) \tag{4.15}$$

where R_p is the M_p transistor channel resistance given by

$$R_p = \frac{1}{\beta_p(V_{DD} + V_{Tp})} = \frac{L}{k'_p W(V_{DD} + V_{Tp})} \tag{4.16}$$

In the same way, the discharge time constant can be estimated as

$$\tau_{\text{dis}} = (R_l + R_n)C_{\text{out}} \tag{4.17}$$

and assuming that both transistors M_l and M_n have the same width W and length L of the channels, we obtain

$$\tau_{\text{dis}} = \frac{2L}{k'_n W(V_{DD} + V_{Tp})} \tag{4.18}$$

Time constants τ_{ch} and τ_{dis} can be estimated on the basis of the given maximum clock frequency f_{max}. Formulae (4.17) and (4.18) can be used to calculate the initial values of the transistor channel dimensions W and L.

A similar analysis can be performed for a pMOS inverter shown in Figure 4.14.

4.2.2 Dynamic Logic Gates

Dynamic digital gates can be realized with the use of the same nMOS and pMOS logic arrays that are used in static logic circuits. Let us compare nMOS and pMOS

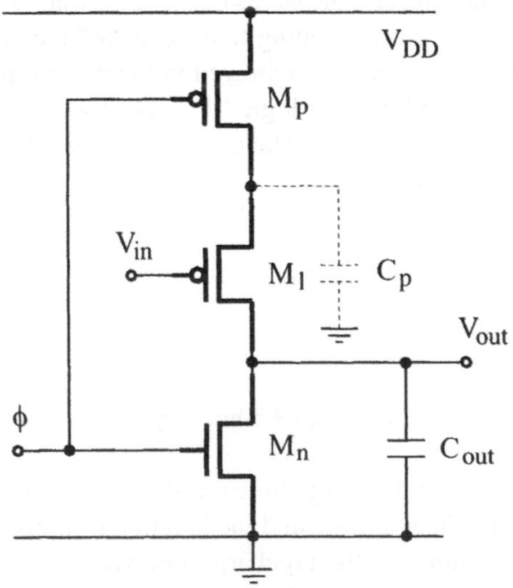

Figure 4.14 Dynamic pMOS inverter.

inverters in Figure 4.13 and Figure 4.14 with the gates in Figure 4.15 a and b. We see that complex dynamic logic gates are obtained if single M_l transistors are replaced by nMOS and pMOS logic arrays.

Dynamic logic gates can be cascaded in a chain composed of digital gates of alternating type, as shown in Figure 4.16. Let us note that the operation of successive stages in different phases of the clock protects the circuit against logic glitches. The precharge and evaluate intervals occur in opposite phases of the clock for the odd and even stages.

a)

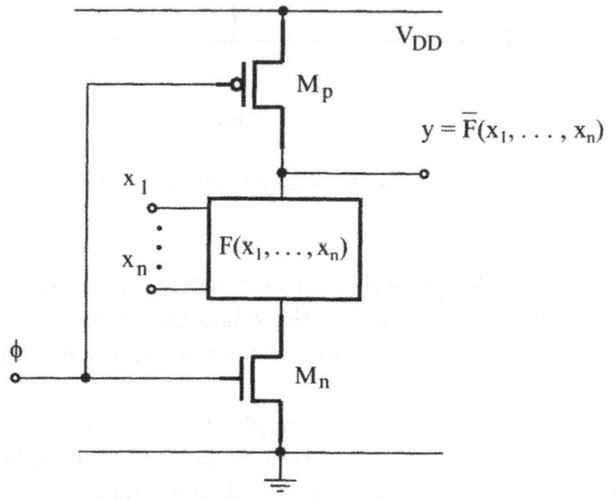

b)

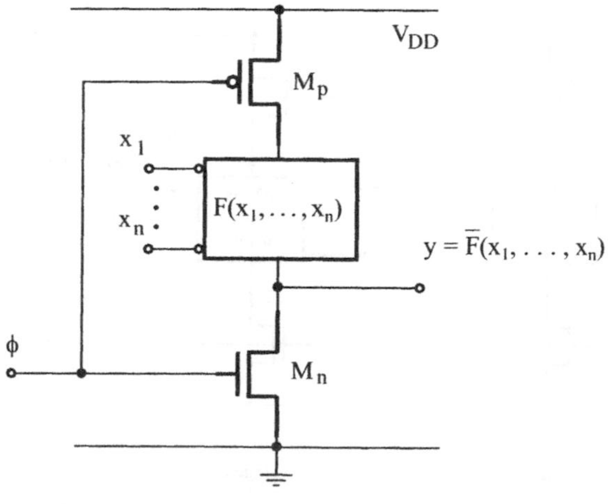

Figure 4.15 Dynamic logic gates.

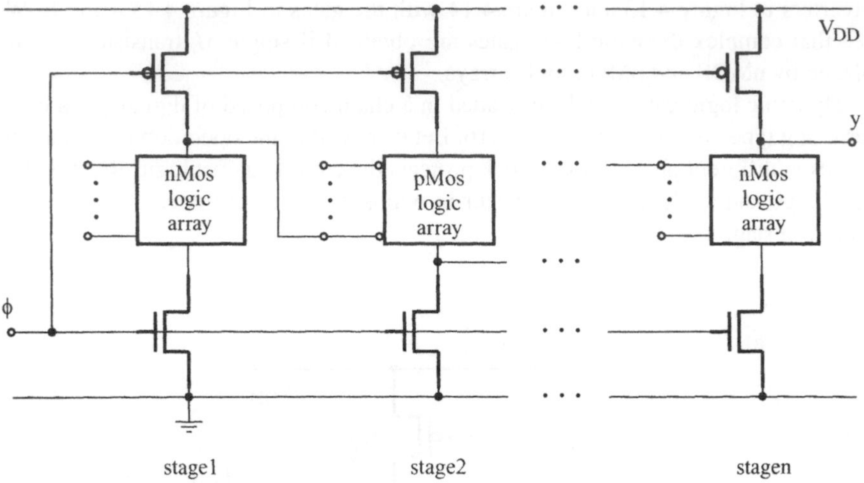

Figure 4.16 Chain of dynamic logic gates.

Another system design style that eliminates logic glitches is the domino logic (DL) shown in Figure 4.17, where the stages are buffered by inverters. The output of the previous stage drives the input of the transistor that is the closest to the output of the next stage.

On the basis of the dynamic nMOS, pMOS logic, a no-race (NORA) logic can be developed that introduces a cascade connection of the Φ-section and $\overline{\Phi}$-section, terminated by a latch. In order to understand signal races, we will consider a cascade connection of transmission gates (TG) and a logic gate with propagation time t_p (Figure 4.18). It is impossible to generate a clock signal as a perfect square wave.

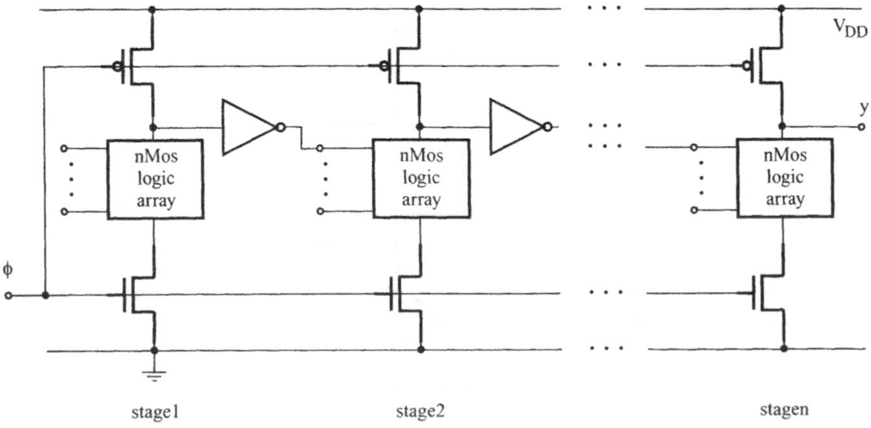

Figure 4.17 Domino logic chain.

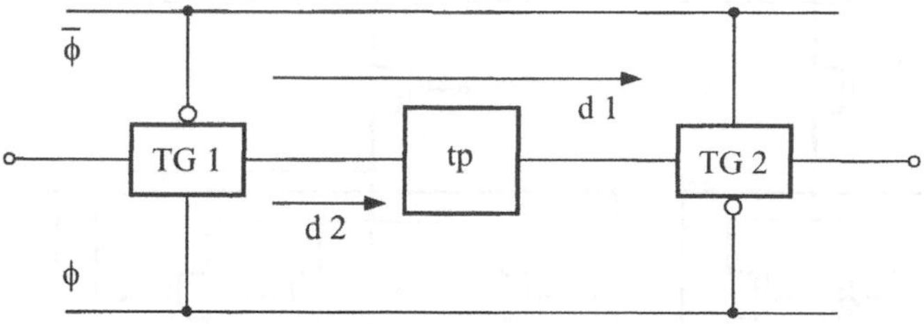

Figure 4.18 Signal race in a logic circuit.

The clock always has finite rise and fall times t_r, t_f. During these transition intervals, both transmission gates, TG_1 and TG_2, can partially conduct. Hence, the data bit d_2 delivered during one clock period can reach the output at the moment when the previous data bit d_1 is still there. We can say that d_2 "wins the race" and d_1 is lost. Such possibility does not exist when the gate propagation time is much greater than the rise and fall times ($t_p \gg t_r, t_f$). The effect of signal race also appears when the clock Φ is delayed with respect to Φ by the skew time $t_{skew} \approx t_p$.

As an example of NORA logic application let us consider the full-adder described by the sum and carry logic functions (4.9) and (4.10). The circuit of the adder is shown in Figure 4.19 and its symbol in Figure 4.20. During the $\Phi = 0$ clock interval, the carry logic function is calculated by a dynamic logic gate containing the pMOS array. During the $\Phi = 1$ clock interval, the sum logic function is calculated in the part of the adder containing the nMOS array. In the $\Phi = 1$ phase, the latch works as an inverter (the nMOS and pMOS transistors with the gates connected to the clock conduct). In the phase $\Phi = 0$, the latch holds the inverted input logic value at its output. The delay unit Δ is composed of two latches. The first one has a carry clear input that resets the carry bit before the LSBs x_{10} and x_{20} arrive.

The described adder, which is an example of a pipelined system, can be used for the realization of a serial–parallel multiplier, which calculates the bits p_j of the product of n-bit figures x_1, x_2 in the form

$$p_j = \sum_{k+l=j} x_{1k}x_{2l}, \qquad k, l = 0, \ldots n-1, \qquad j = 0, \ldots, m, \qquad m = 2(n-1) \tag{4.19}$$

The above formula for four-bit words x_1, x_2 can be written in the form

$$\begin{array}{llllll}
p_0 = & & & & x_{10}x_{20} \\
p_1 = & & & x_{10}x_{21} & +x_{11}x_{20} \\
p_2 = & & x_{10}x_{22} & +x_{11}x_{21} & +x_{12}x_{20} \\
p_3 = & x_{10}x_{23} & +x_{11}x_{22} & +x_{12}x_{21} & +x_{13}x_{20} \\
p_4 = & x_{11}x_{23} & +x_{12}x_{22} & +x_{13}x_{21} & \\
p_5 = & x_{12}x_{23} & +x_{13}x_{22} & & \\
p_6 = & x_{13}x_{23} & & &
\end{array} \tag{4.20}$$

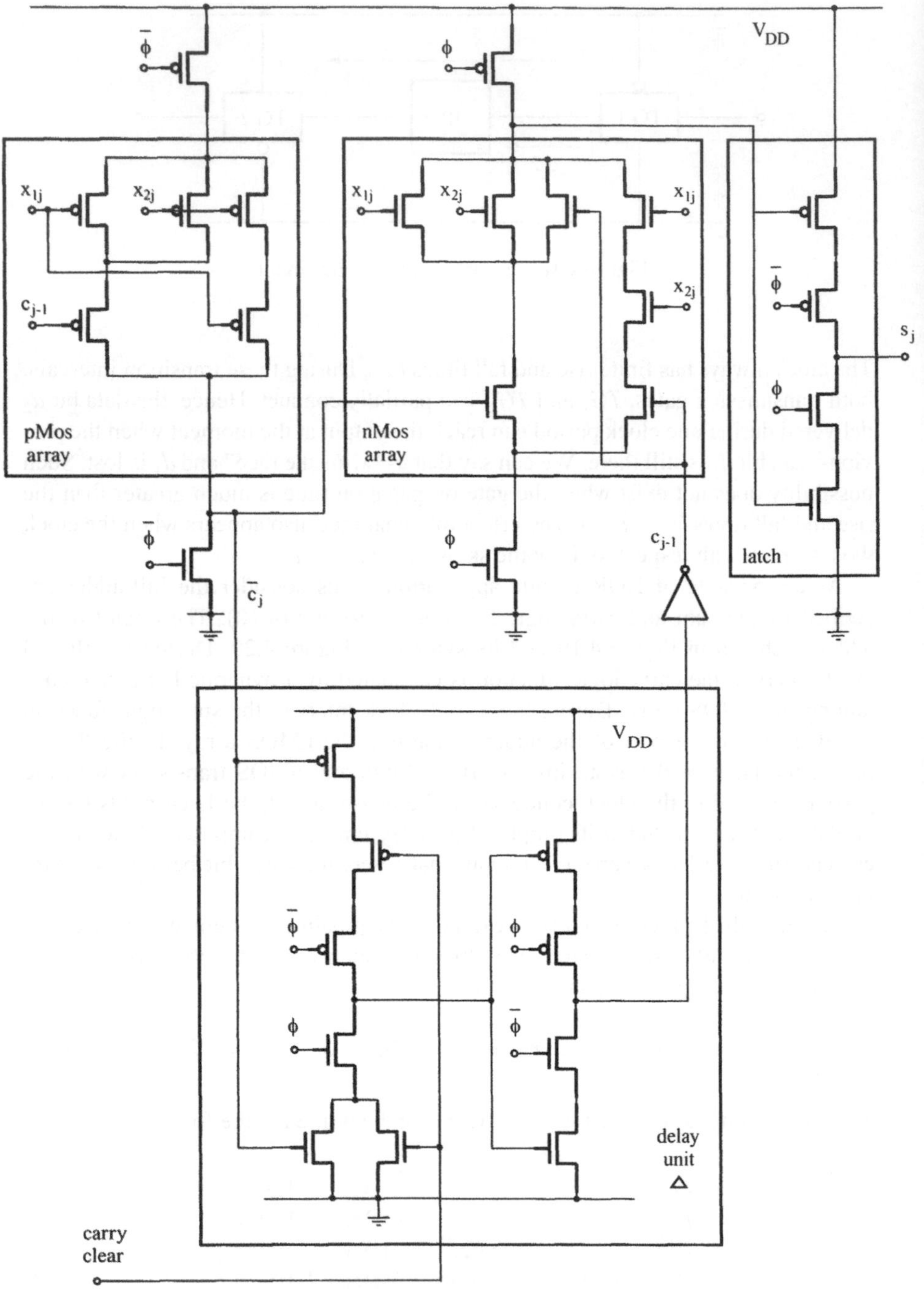

Figure 4.19 NORA serial full-adder.

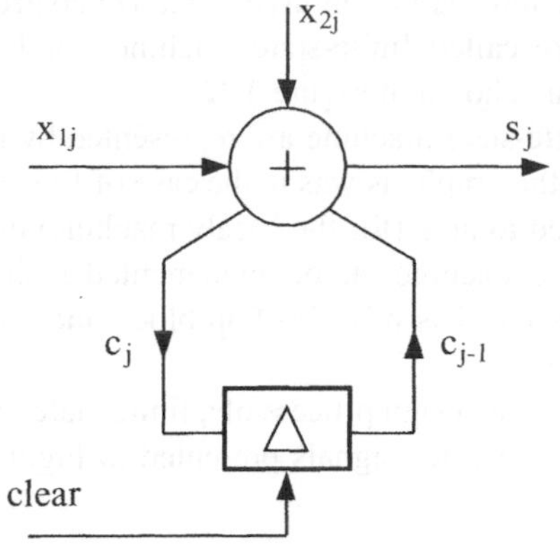

Figure 4.20 NORA serial full-adder symbol.

and implemented as the four-bit serial–parallel multiplier shown in Figure 4.21. We see that in this implementation, all bits of x_1 are multiplied by each bit of x_2 and the results are then added with the use of four serial adders to obtain the bits p_j, $j = 0, \ldots, m - 1$ of the product.

4.3 FINITE-STATE MACHINES

Flip-flops, introduced in the previous sections, were described with the use of graphs (state-transition diagrams). These graphs contained only two nodes. Howev-

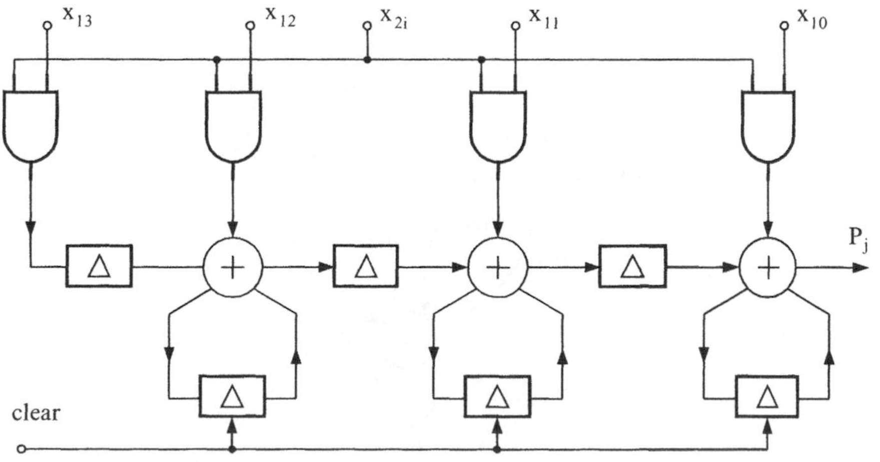

Figure 4.21 Serial–parallel multiplier.

er, more complex structures are often used to realize controllers of simple processes. Such structures are called finite-state machines, or FSMs. Two examples of three-state machines are shown in Figure 4.22.

The states of a finite-state machine are represented by nodes and input signals, $x_1, \ldots, x_k$, by arcs of the graph, as was in the case of flip-flops. The output signals, $y_1, \ldots, y_l$, are assigned to arcs (for the Mealy machine) or to nodes (for the More machine). A finite-state machine can be implemented as the network shown in Figure 4.23. The network consists of a flip-flop block and combinational logic block composed of logic gates.

In order to illustrate the design process of a finite-state machine, let us consider a four-phase clock that generates signals presented in Figure 4.24. Such multiphase

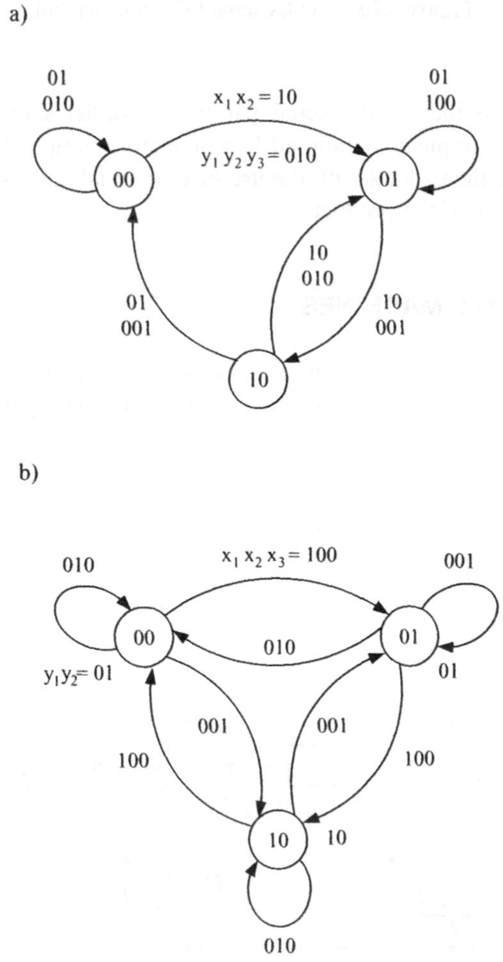

Figure 4.22 Examples of Mealy (a) and Moore (b) finite-state machines.

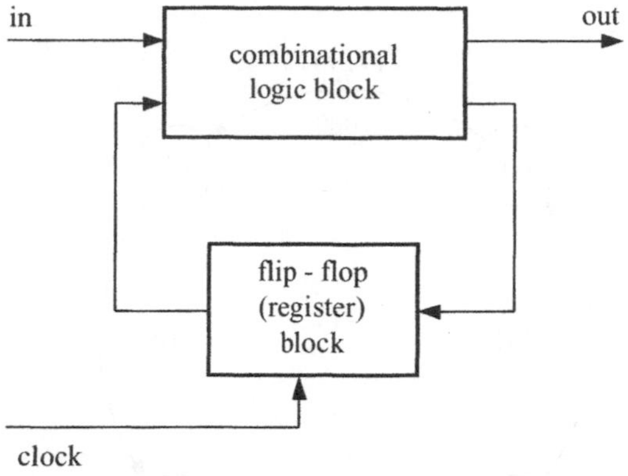

Figure 4.23 Finite-state machine implemented with the use of a flip-flop block and a combinational logic block.

clock generators are often used to control circuits implemented in switched capacitor or switched current techniques. The state diagram of the Mealy machine, which is now an autonomous FSM (input signals do not occur and are not assigned to the arcs), is shown in Figure 4.25a.

The design procedure is as follows. Each arc is described in one row of the truth Table 4.2. The first column contains bits of the current states, the second one of the next states, and the third one of the outputs. In the general case, when input signals

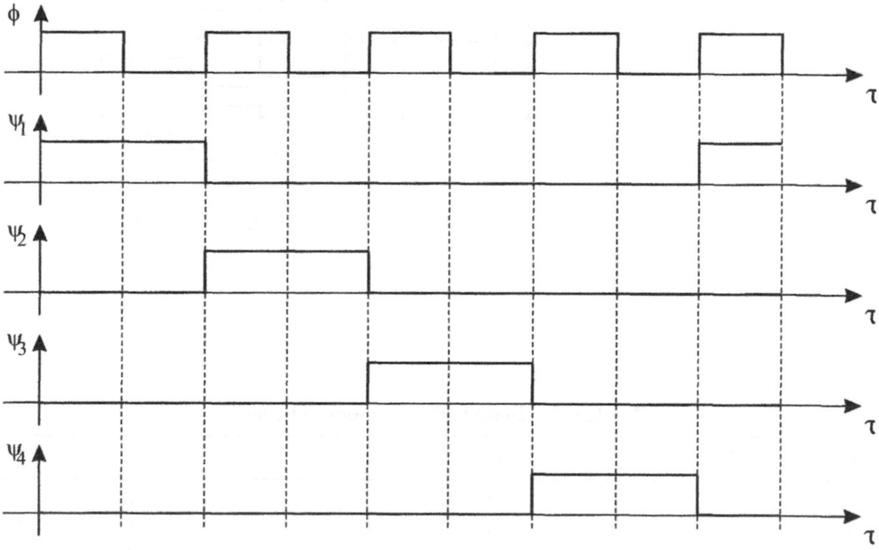

Figure 4.24 Four-phase clock waveforms.

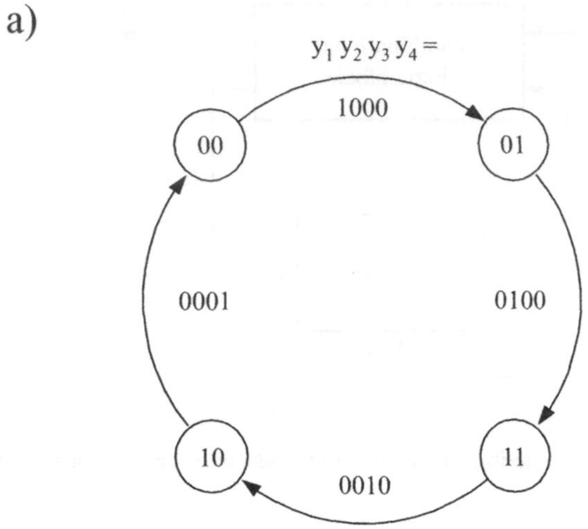

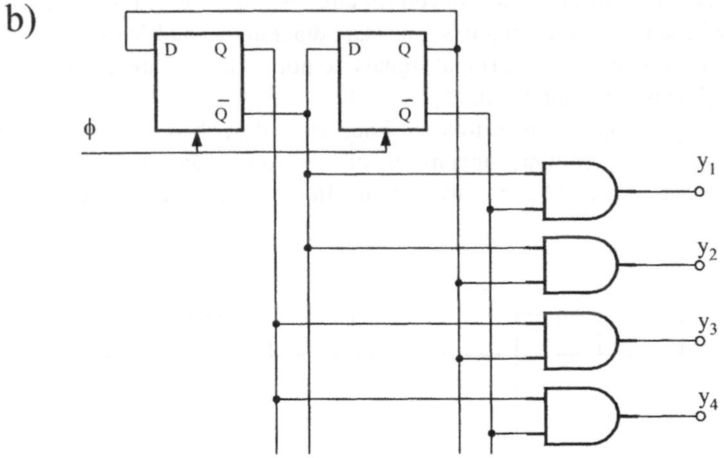

Figure 4.25 Four-phase clock generator: the state transition diagram (a), and implementation (b).

Table 4.2 State table of a clock generator

Q_1	Q_2	Q'_1	Q'_2	y_1	y_2	y_3	y_4
0	0	0	1	1	0	0	0
0	1	1	1	0	1	0	0
1	1	1	0	0	0	1	0
1	0	0	0	0	0	0	1

are represented in the FSM, they are included in the first column of the table, too. From this table we obtain the logic equations of the bits of the next states in the form

$$
\begin{aligned}
Q_1' &= \overline{Q}_1 Q_2 + Q_1 Q_2 = Q_2, \\
Q_2' &= \overline{Q}_1 \overline{Q}_2 + Q_1 Q_2 = Q_1
\end{aligned} \qquad (4.21)
$$

These logic equations result in the connections of the DFF shown in Figure 4.25b. The third column of the table implicates the logic equations

$$
\begin{aligned}
y_1 &= \overline{Q}_1 \overline{Q}_2, \\
y_2 &= \overline{Q}_1 Q_2, \\
y_3 &= Q_1 Q_2, \\
y_4 &= Q_1 \overline{Q}_2,
\end{aligned} \qquad (4.22)
$$

which are implemented with the use of AND gates in the FSM in Figure 4.25b.

4.4 MEMORIES

This section briefly describes digital memories that can be realized in the CMOS technology. A memory can be considered as a matrix of cells (Figure 4.26). Each

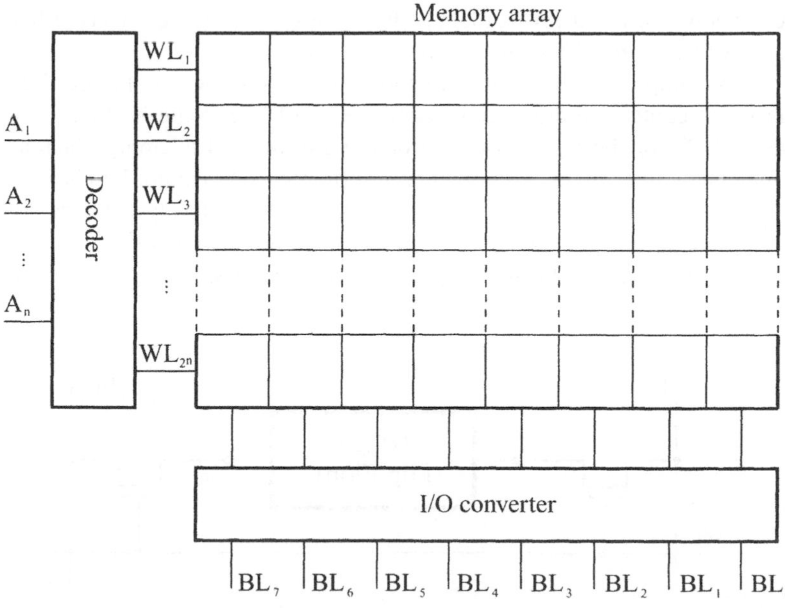

Figure 4.26 Architecture of a digital memory.

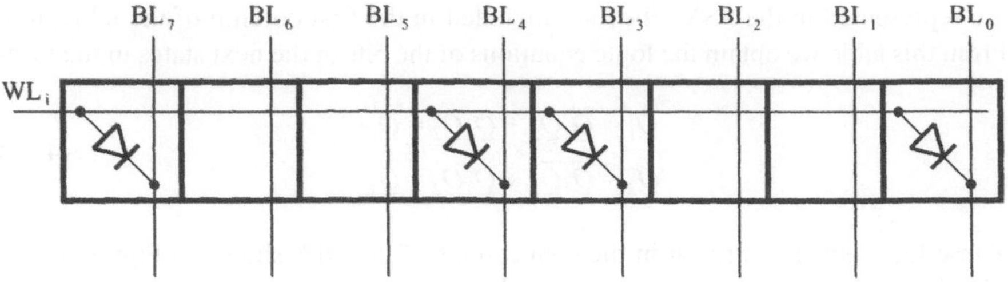

Figure 4.27 Word implementation of memory based on diodes.

cell stores one bit of digital data. The memory shown in Figure 4.26 has eight cells in each row of the memory array. The data stored in one row is called a word. The memory shown in Figure 4.26 stores eight-bit (one byte) words. The abbreviation for byte is the capital letter B, hence the word size in our example can be denoted as 8b = 1B. Note that *word* is not a synonym of *byte*.

In order to read or write a word from or into the memory, the cells in columns of the memory array are connected to an I/O converter by bit lines (BLs). A set of such wires that together carry a binary number is called a bus. The I/O converter can be composed of resistors or more complicated elements like amplifiers and buffers, depending on the kind of memory. The address delivered to the address bus $A_1, \ldots, A_n$ points at the chosen word of the memory. For this purpose, the decoder, composed of digital gates and multiplexers (Figure 4.26), converts an n-bit address into a logic "1" delivered to one of 2^n word lines (WLs), leaving "0" on the remaining word lines. The memory size depends on the word size as well as on the number of address bus wires. For a 1B word and address bus $n = 10$ we get 1 kB (1024 B) of memory, while for $n = 16$ we get 64 kB of memory.

A simple implementation of a memory word is shown in Figure 4.27. The memory cell contains a diode connecting the word line to the bit line in the case of the bit "1." For the bit "0" of the stored word, the cell is empty. Such a memory implementation contains permanent information that cannot be changed. It is called read-only memory or ROM. A programmable read-only memory (PROM) can be obtained for the memory array completely filled with diodes. Selected diodes are disconnected during the programming process. If such connections can

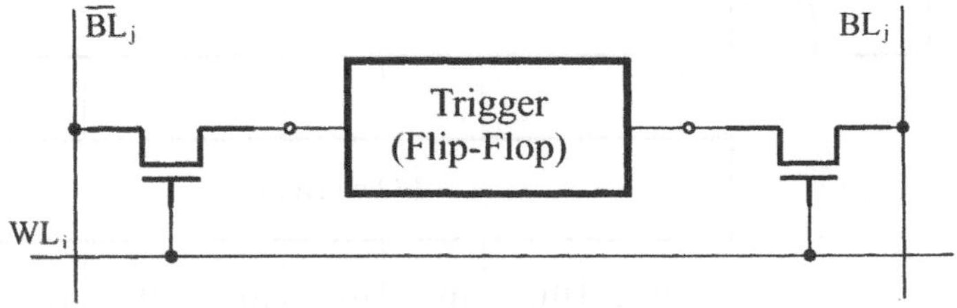

Figure 4.28 Flip-flop memory cell.

be changed and if the programming process can be repeated, then the memory is called erasable programmable read-only memory (EPROM). In this case, the diodes are replaced by MOS transistors that act either as diodes or as open circuits, depending on whether their gates are electrically charged or not. An electrically erasable programmable read only memory (EEPROM) is often called a FLASH memory.

A disadvantage of the read-only memory is that its contents cannot be changed or can be changed by a relatively long programming process. The memory com-

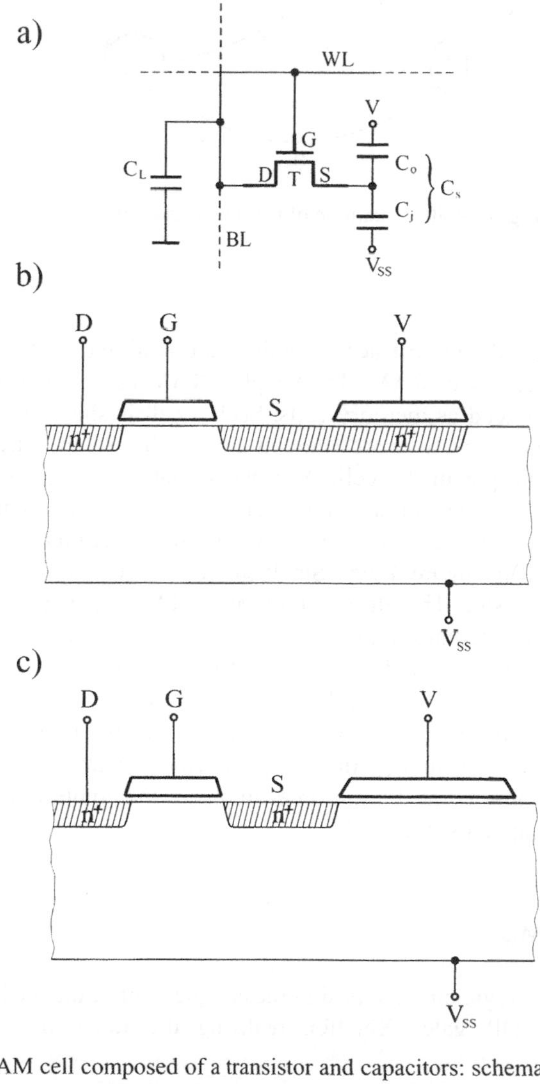

Figure 4.29 RAM cell composed of a transistor and capacitors: schematic diagram (a) and its implementations in Dennard (b) and Kosonocky (c) structures.

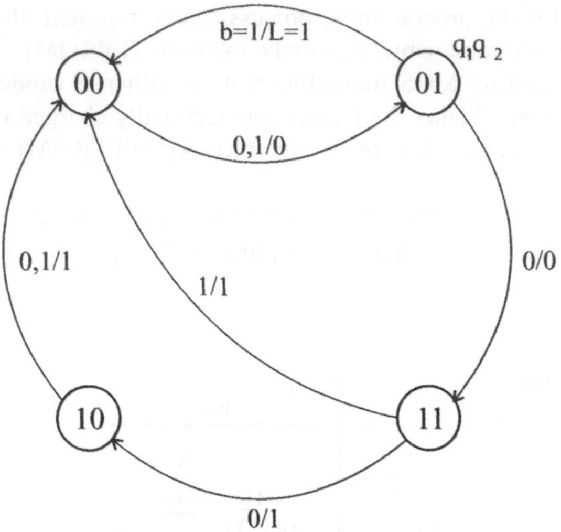

Figure 4.30 Example of the state transition diagram.

posed of cells that allows fast access both in the read and write modes is called random access memory or RAM. The simplest RAM can be realized with the use of flip-flops introduced as memory cells. Such a cell is shown in Figure 4.28 together with connections to word and bit lines. The Schmitt trigger shown in Figure 2.19 can be put in the cell. A memory that uses triggers to hold bits is called static RAM. Large chip area and high power consumption are disadvantage of static cells. The dynamic RAM cell in which tiny capacitors are used as storage elements is shown in Figure 4.29a. Small stored charges are quickly lost through the connected transistor. Therefore, the dynamic RAM needs a refreshing process before the charge becomes unreadable. Two implementations of the dynamic RAM cell, called Dennard and Kosonocky structures, are shown in Figure 4.29 b and c, respectively. In the Dennard structure the common plate of capacitors C_0 and C_j is obtained by extension of the source area of the transistor. In this case, the node V is connected to V_{SS}. In the Kosonocky structure, the node V must be connected to the power line V_{DD} in order to obtain the common plate and to realize the junction capacitor C_j.

4.5 PROBLEMS

1. Using static logic circuits, design the complement of the XOR element called exclusive-NOR gate (XNOR), realizing the function $F = x_1 \odot x_2 = \overline{x_1 \oplus x_2}$.

2. Using the gate whose general structure is presented in Figure 4.15a, realize a half-adder whose sum and carry functions are described by relations (4.8)
3. The state transition diagram in Figure 4.30 describes a two-speed flasher. The flasher emits light only when the output signal $L = 1$. For the button b on ($b = 1$), the oscillations of flashes are twice as fast as for the button b off ($b = 0$). Realize this FSM.
4. Explain why the diode connections in Figure 4.27 cannot be replaced by short-circuit connections.

5

Analog VLSI Circuits

In this chapter, we present the basic elements of passive and active circuits. The simplest one-port elements are the resistance, the inductance, and the capacitance. The more complex two-port elements are the ideal gyrator, the ideal transformer, the transducers, and the nullor. We describe various implementations of integrators, i.e., switched-current, switched-capacitor, and OTA-C implementations. These integrators are basic cells of mixed signal processing systems. We also consider low-voltage and low-power operation of these elements.

5.1 PASSIVE AND ACTIVE CIRCUIT ELEMENTS

Passivity is a very important property of circuits. It ensures stability and low noise. However, it is difficult to integrate circuits composed mainly of resistors, capacitors, and inductors in systems with high-precision signal processing. They also introduce attenuation. On the other hand, CMOS technology offers as basic elements n- and p-MOS transistors. Hence, we are interested in active elements that can be built from transistors.

A passive element is theoretically defined as an element that consumes nonnegative energy:

$$E(t) = \int_{-\infty}^{t} p(\tau)d\tau \ 0, \qquad \text{for all } t, \qquad E(-\infty) = 0, \qquad (5.1)$$

where $p(t)$ is the power delivered to all ports of an element. In one-port elements, the resistance $R[v(t) = Ri(t)]$, the inductance $L\{v(t) = L[di(t)/dt]\}$, and the capacitance $C\{i(t) = C[dv(t)/dt]\}$, are passive elements. If the energy in definition (5.1) is equal to zero $[E(t) = 0]$, then we speak about lossless elements.

The ideal transformer (IT) and the ideal gyrator (IG) [41] are examples of a passive two-port network presented in Figure 5.1. The input impedance Z_{in} denotes the impedance seen at the primary port, with the secondary port loaded by Z_L. The ideal transformer is a typical example of a converter; the ideal gyrator is a typical example of an inverter.

The two-port network loaded on one of its ports by impedance Z_L and with the input impedance Z_{in}, seen from the other port, directly (inversely) proportional to the load impedance

$$Z_{in} = k_c Z_L, \qquad (Z_{in} = k_i/Z_L) \tag{5.2}$$

is called a converter (inverter).

It follows from the definition above that the chain matrix of the converter has nonzero elements only on the diagonal, whereas the chain matrix of the inverter has elements on the diagonal equal to zero. Hence, the chain matrices of the transformer and the gyrator are as follows:

$$F_{IT} = \begin{bmatrix} 1/n & 0 \\ 0 & n \end{bmatrix} \tag{5.3}$$

and

$$F_{IG} = \begin{bmatrix} 0 & 1/g \\ g & 0 \end{bmatrix} \tag{5.4}$$

Matrix (5.4) implies that $gv_1 = i_2' = -i_2$, $i_1 = gv_2$, hence the total power delivered to both ports of the gyrator is equal to zero:

$$gv_1 i_1 + gv_2 i_2 = g[p_1(t) + p_2(t)] = 0 \tag{5.5}$$

According to definition (5.1) the ideal gyrator is a lossless element.

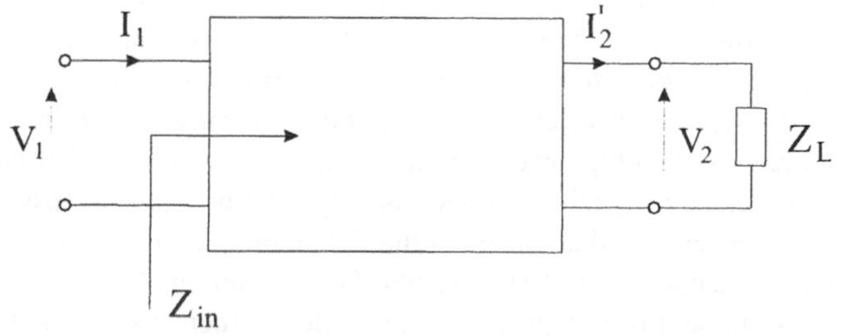

Figure 5.1 Two-port network loaded by impedance Z_L.

In chain matrices of a converter and an inverter, two elements are equal to zero. In comparison, chain matrices of transducers (dependent sources) have three elements equal to zero. Transducers are the most popular active elements. We will call two-port networks defined by the chain matrices

$$F_{VVT}=\begin{bmatrix}1/\mu & 0\\ 0 & 0\end{bmatrix},\quad F_{CCT}=\begin{bmatrix}0 & 0\\ 0 & 1/\alpha\end{bmatrix}$$

$$F_{VCT}=\begin{bmatrix}0 & 1/g\\ 0 & 0\end{bmatrix},\quad F_{CVT}=\begin{bmatrix}0 & 0\\ 1/r & 0\end{bmatrix} \tag{5.6}$$

voltage to voltage (VVT), current to current (CCT), voltage to current (VCT), and current to voltage (CVT) transducers.

Another active element, with very important implementations, is a nullor. All the elements of its chain matrix are equal to zero:

$$F=0 \tag{5.7}$$

The chain matrix of the nullor implies that the input voltage and current are equal to zero ($V_1 = 0$, $I_1 = 0$), and that V_2, I'_2 at the output are indefinite signals. The most popular representations of the nullor are an ideal operational amplifier (op-amp) and an ideal operational transconductance amplifier (OTA). The ideal operational amplifier is characterized by the input impedance tending to infinity and the output impedance equal to zero, and the ideal operational transconductance amplifier (OTA) is characterized by both (input and output) impedances tending to infinity. The CMOS realizations of op-amp and OTA circuits are both based on a differential input stage, as shown in Figure 2.29, Figure 2.27, and Figure 2.28.

5.2 OTA-C CIRCUITS

OTA-C circuits are composed of operational transconductance amlifiers (OTAs) and capacitors. Design parameters of a circuit are transconductances g of amplifiers and capacitances C of capacitors. Both kinds of parameters are realized on a chip with different tolerances. The capacitance tolerance is primarily determined by the variation of the dielectric thickness, whereas the transconductance tolerance is determined by the variations of channel dimensions of transistors. Hence, OTA-C circuits do not have such a high precision in signal processing as SC circuits, in which design parameters are capacitances, exclusively. The processing based on the OTA-C technique is faster than that based on the SC technique, in which a clock is necessary to control the operation of SC circuits. The continuous-time operation of OTA-C circuits is an additional property that allows their use as antialiasing and smoothing filters in the system shown in Figure 1.1.

5.3 SWITCHED CURRENT TECHNIQUE

In this section, we will describe a recently introduced technique, the so-called switched current (SI) technique. In comparison with Ota-C circuits, the SI circuits operate with discrete-time signals and can be realized together with digital circuits on the same chip, with no additional technological processes.

5.3.1 Components of SI Circuits

The main property of an SI circuit is its operation in the current mode. Each linear discrete-time system can be realized with the use of three basic elements: summer, multiplier (scaling element), and delay element. Because of the current mode operation of SI circuits, the first element is realized simply as a node in which current signals are summarized. The current mirror presented in Figure 5.2a can be used as a scaling element.

Delay elements are realized with the use of memory cells that are the basic components of SI circuits. The so-called first generation memory cell is obtained on the basis of the current mirror, by introducing one switch, as shown in Figure 5.2b.

The second-generation memory cell, presented in Figure 5.3a, is composed of switches, the bias transistor M_p, and the memory transistor M_n. Switches are con-

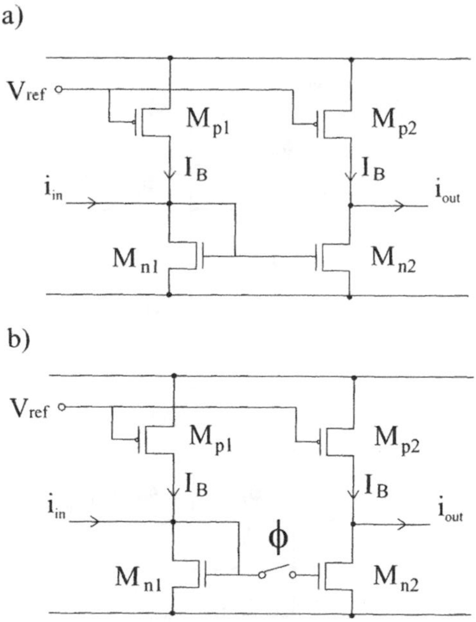

Figure 5.2 Current mirror used as a multiplier in SI circuits (a) and the first-generation memory cell obtained on the basis of a current mirror (b).

trolled by a two-phase clock and are periodically closed and opened in opposite phases Φ_1, Φ_2. The cell works in read and write modes. In the write mode, when the switches controlled by Φ_1 are closed, the input current $i_{in}(nT)$ in the n-th clock period T is added to the bias current I_B, giving the drain current of the diode-connected memory transistor M_n. Hence, this transistor plays the role of the transistor M_{n1} of the first-generation memory cell. In the read mode, when the switch controlled by Φ_2 is closed, the gate voltage is held on the gate capacitor, giving the same value of the drain current as in the write mode. Hence, in order to meet the Kirchhoff's current law (KCL), the output current $i_{out}(nT + T/2) = -i_{in}(nT)$. We see that the same transistor, M_{n1}, now plays the role of the transistor M_{n2} of the first-generation memory cell.

In order to minimize parasitic effects in the second-generation memory cell, various cell structures have been developed. The balanced structure is used in order to

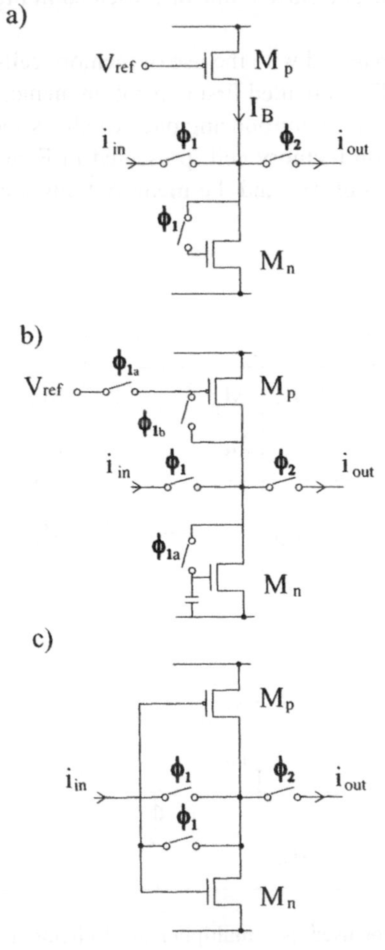

Figure 5.3 Second-generation memory cells.

cancel dc offset and clock feed-through error; the grounded-gate feedback structure is used for minimizing a transmission error in cascaded memory cells, etc. A very interesting structure is presented in Figure 5.3b. In this so-called S^2I memory cell [27], the period of write mode is divided into two steps, *a* and *b*, during which the coarse and then the fine memorizing occurs. Both the nMOS and pMOS transistors serve as memory transistors, the first one in phase Φ_{1a}, the second one in Φ_{1b}. The S^2I memory cell needs twice the clock frequency to control the switches as the second-generation memory cell.

A simple and effective memory cell is presented in Figure 5.3c [24, 23]. During phase Φ_1, the nMOS and pMOS complementary transistors are diode-connected, and thanks to careful design approximately one-half ($i_{in}/2$) of the input current i_{in} is memorized in each transistor.

In order to analyze this memory cell in detail, let us consider the complementary pair of transistors presented in Figure 5.4. Assuming that both transistors work in saturation regions, the drain currents can be written from (2.5) and (2.8) as:

$$I_{Dp} = \frac{\beta_p}{2}(V_{gp} - V_{DD} - V_{Tp})^2, I_{Dn} = \frac{\beta_n}{2}(V_{gn} - V_{SS} - V_{Tn})^2 \qquad (5.8)$$

Assuming

$$V_{DD} = -V_{SS} = V_o, V_{gp} = V_{gn} = V_g \qquad (5.9)$$

we obtain

$$\begin{aligned} I = I_{Dn} - I_{Dp} &= \tfrac{1}{2}(\beta_n - \beta_p)V_g^2 \\ &+ [\beta_n(V_o - V_{Tn}) + \beta_p(V_o + V_{Tp})]V_g + \tfrac{1}{2}[\beta_n(V_o - V_{Tn})^2 - \beta_p(V_o + V_{Tp})^2] \end{aligned} \qquad (5.10)$$

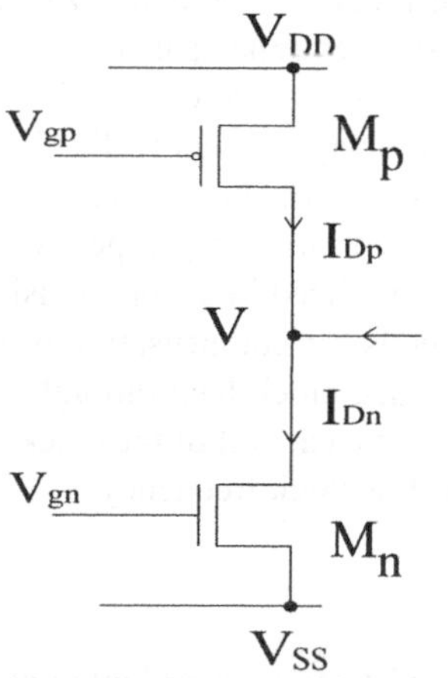

Figure 5.4 Complementary pair of nMOS and pMOS transistors.

The first component in equation (5.10) describes the distortion of the complementary pair. The second component describes the linear dependence of the current I with respect to the voltage V_g, where the transconductance is given by $[\beta_n(V_o - V_{Tn}) + \beta_p(V_o + V_{Tp})]$. The third component denotes the dc offset of the cell. The dc offset can be effectively cancelled when circuits are realized in a balanced structure. The distortion can be cancelled for $\beta_n = \beta_p$ or $k'_n W_n/L_n = k'_p W_p/L_p$. Let us note that for $\beta_n = \beta_p$, nearly the same levels of the voltages at the input and the output of the complementary pair can be achieved. This property is very important for the minimization of transmission error in cascaded cells. An example of cascade connection will be shown in Figure 5.19 and later considered in one of the problems.

Assumptions (5.9) correspond to connections realized in the circuit in Figure 5.3c. For the circuit in Figure 5.3a, adequate assumptions are

$$V_{DD} = -V_{SS} = V_o, \qquad V_{gp} = V_{\text{ref}}, \qquad V_{gn} = V_g \tag{5.11}$$

for which we obtain:

$$I = I_{Dn} - I_{Dp} = \frac{\beta_n}{2} V_g^2 + \beta_n(V_o - V_{Tn})V_g$$
$$- \frac{\beta_p}{2} V_{\text{ref}}^2 + \beta_p(V_o + V_{Tp})V_{\text{ref}} + \frac{\beta_n}{2}(V_o - V_{Tn})^2 - \frac{\beta_p}{2}(V_o + V_{Tp})^2 \tag{5.12}$$

The comparison of equations (5.10) and (5.12) shows that in the memory cell in Figure 5.3a the distortion is greater and the transconductance smaller than in the memory cell in Figure 5.3c.

The cell presented in Figure 5.3c is comparable to the high-performance cell presented in Figure 5.3b with respect to current error occurring due to charge injection. Simultaneously, the dynamic range of the memory cell in Figure 5.3c is twice as great, because of the previously mentioned partition of the input current i_{in} into two branches, increasing the signal-to-noise ratio by 6 dB.

The schematic diagram of a delay element realized with the use of memory cells shown in Figure 5.3a, b, or c is presented in Figure 5.5 [28]. The blocks P and N denote pMOS and nMOS memory transistors, respectively, with switches connected to their gates. The delay cell is realized in the previously mentioned balanced structure, which is very advantageous for compensation of parasitic effects like dc offset, crosstalk from digital signals, clock feed-through, and charge injection. Let us note that the signal is sampled in each half of the clock period. Hence, the sampling frequency is twice as great as the clock frequency.

5.3.2 Integrators

Memory cells are basic parts of delay lines and integrators, which are indispensable in signal and image processing performed in the switched current technique. In this

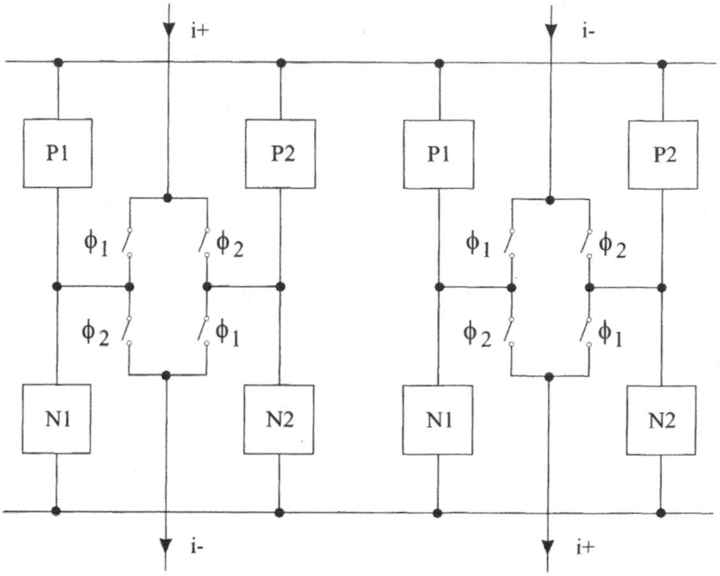

Figure 5.5 Schematic diagram of a delay element.

section, we consider two structures of bilinear balanced integrators. The first integrator is composed of inverting and noninverting Euler integrators [28]. The basic structure of noninverting and inverting Euler integrators, based on the memory cell from Figure 5.3a, is presented in Figurc 5.6.

In order to calculate the transfer function of the noninverting integrator, let us consider the second phase of the $n-1$ clock period. In this phase, the switches controlled by Φ_2 are closed, the first transistor M_{n1} is diode connected, and its drain current can be calculated as:

$$\begin{aligned} i_{D1} &= 2I_B + i_{\text{in}}(n-1) - i_{D2} = 2I_B + i_{\text{in}}(n-1) - [I_B - i_{\text{out}}(n-1)] \\ &= I_B + i_{\text{in}}(n-1) + i_{\text{out}}(n-1) \end{aligned} \tag{5.13}$$

In the first phase of the nth clock period, the switches controlled by Φ_2 are open, the drain current i_{D1} is held and the drain current of the second transistor M_{n2} is as follows:

$$i_{D2} = 2I_B - i_{D1} = I_B - i_{\text{in}}(n-1) - i_{\text{out}}(n-1) \tag{5.14}$$

In this phase, the switch controlled by Φ_1 is closed and the transistors M_{n2} and M_{n3} work as a current mirror, giving

$$i_{\text{out}}(n) = I_B - i_{D2} = i_{\text{in}}(n-1) + i_{\text{out}}(n-1) \tag{5.15}$$

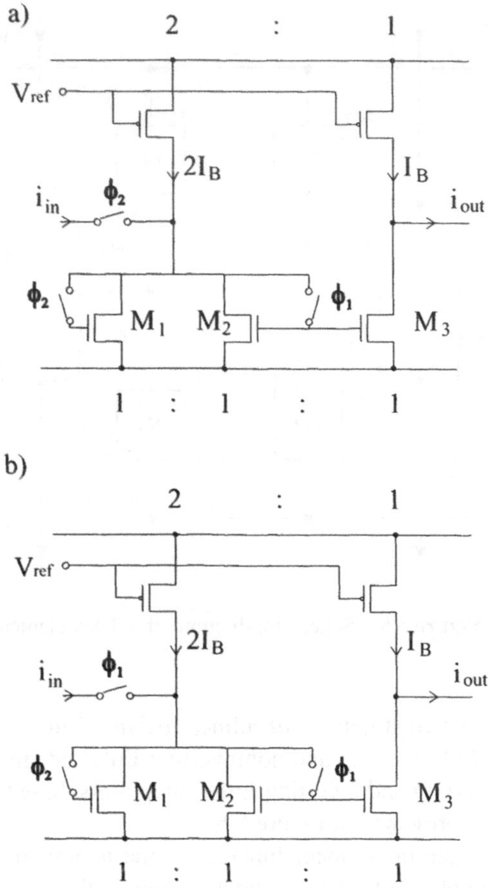

Figure 5.6 Noninverting (a) and inverting (b) Euler integrators.

Hence, the transfer function $H(z)$ of the integrator is as follows

$$H(z) = \frac{I_{out}(z)}{I_{in}(z)} = \frac{z^{-1}}{1 - z^{-1}} \tag{5.16}$$

The comparison of the above transfer function with the transfer function

$$H(s) = \frac{V_{out}(s)}{V_{in}(s)} = \frac{-G}{sC} \tag{5.17}$$

of an integrator composed of an op-amp, the input branch characterized by conductance G and the feedback branch characterized by capacitance C, shows that for $GT/C = 1$ the noninverting Euler integrator (5.16) is a discrete counterpart network of an analog integrator realizing the forward Euler transformation, $sT = z - 1$ (1.26).

By analogy, the drain current of the first transistor M_{n1}, in the second phase of the $n-1$th clock period of the inverting integrator in Figure 5.6b, can be calculated as follows:

$$i_{D1} = 2I_B - i_{D2} = 2I_B - [I_B - i_{out}(n-1)] = I_B + i_{out}(n-1) \tag{5.18}$$

In the first phase of the nth clock period, the drain current of the second transistor M_{n2} is as follows:

$$i_{D2} = 2I_B + i_{in}(n) - i_{D1} = I_B - i_{in}(n) - i_{out}(n-1) \tag{5.19}$$

and the output current is

$$i_{out}(n) = I_B - i_{D2} = -i_{in}(n) + i_{out}(n-1) \tag{5.20}$$

Hence, we get the following transfer function $H(z)$ of the integrator

$$H(z) = \frac{I_{out}(z)}{I_{in}(z)} = \frac{-1}{1 - z^{-1}} \tag{5.21}$$

This integrator is a discrete counterpart network of the integrator (5.17) in which the backward Euler transformation, $sT = 1 - z^{-1}$ (1.27), is realized.

If the noniverting Euler integrator is excited by the signal $I(z)$ and the inverting one by the inverted signal $-I(z)$, then the summated output signals of both integrators give the response of the bilinear integrator. The schematic realization of this kind of bilinear integrator is presented in Figure 5.7 [28]. For $GT/2C = 1$ the integrator is a discrete counterpart network of the integrator (5.17) that realizes bilinear transformation (1.28).

The second kind of bilinear integrator, realized in a structure that we will call the classic one, is presented in Figure 5.8. In order to replicate input and output signals, the multioutput current mirrors with the gains α and β, respectively, are used. The symbol δz^{-1} denotes the balanced delay cell with the gain error δ, described in the previous subsection and presented in Figure 5.5. The transfer function of this cell is described by the formula

$$H_i(z) = \beta_i \frac{1 + \alpha_0 \delta / \alpha_1 z^{-1}}{1/\alpha_1 - \beta_0 \delta / \alpha_1 z^{-1}} \tag{5.22}$$

Such an integrator can be a discrete counterpart network of an analog damping integrator with the transfer function

$$H_{ai}(s) = \frac{h_i}{1 + s\tau} = \frac{a_i}{s + \alpha} \tag{5.23}$$

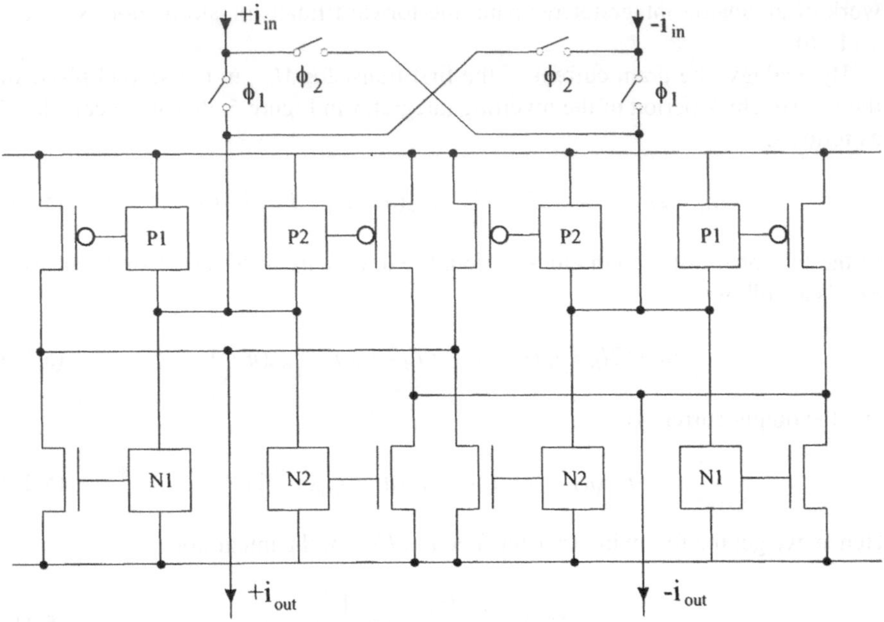

Figure 5.7 Bilinear integrator composed of noninverting and inverting Euler integrators.

where the damping factor $\alpha = 1/\tau$. Assuming that $\tau > T/2$, and comparing the transfer functions (5.22) and (5.23) with the use of bilinear transformation (1.28) between discrete z and analog s domains, we obtain

$$\alpha_1 = \frac{1}{2\tau/T + 1}, \qquad \alpha_0 = \frac{1}{\delta(2\tau/T + 1)} \tag{5.24}$$

and

$$\beta_i = h_i, \qquad \beta_0 = \frac{1}{\delta}\,\frac{2\tau/T - 1}{2\tau/T + 1} \tag{5.25}$$

for given h_i, τ, T, and δ, or

$$\alpha_1 = \frac{1}{2/T + \alpha}, \qquad \alpha_0 = \frac{1}{\delta\,(2/T + \alpha)} \tag{5.26}$$

and

$$\beta_i = a_i, \qquad \beta_0 = \frac{1}{\delta}\,\frac{2/T - \alpha}{2/T + \alpha} \tag{5.27}$$

for given a_i, α, T, and δ.

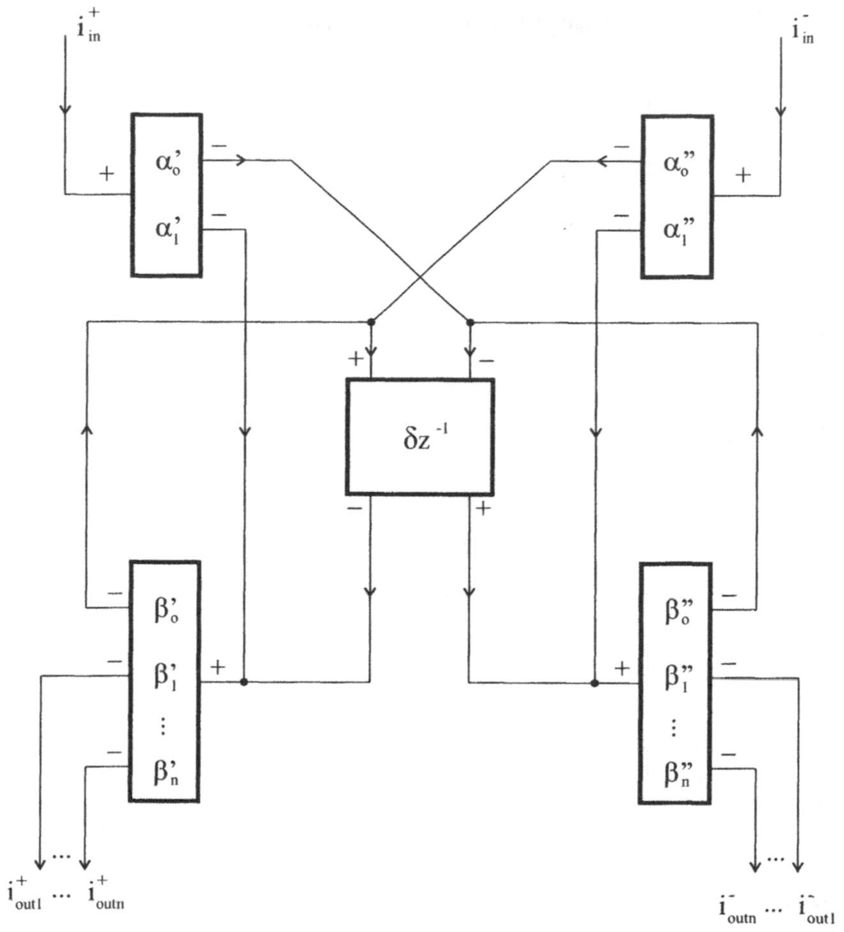

Figure 5.8 Classic structure of bilinear damped integrator.

Examples of impulse responses of balanced integrators are shown in Figures 5.9, 5.10, and 5.11. Technology parameters of the AMS cyx 0.8 CMOS process were used in HSPICE simulations based on the model of the level 49. Dimensions of channels were 56/4 for nMOS and 120/4 for pMOS transistors. We observe that the parasitic effect compensation in the classic integrator is better than in the integrator based on Euler integrators. The large range of damping coefficient regulation is another valuable property of the classic integrator. The waveform in Figure 5.11 shows the impulse response of this integrator in which the gain error δ of the delay cell is cancelled by scaling coefficients α_2 and β_2 of current mirrors. The sampling frequency 20 MHz is achieved in these simulations, despite extremely large transistor dimensions. Other simulations also showed proper operation of the integrator composed of transistors with dimensions reduced to a minimum in the considered technology.

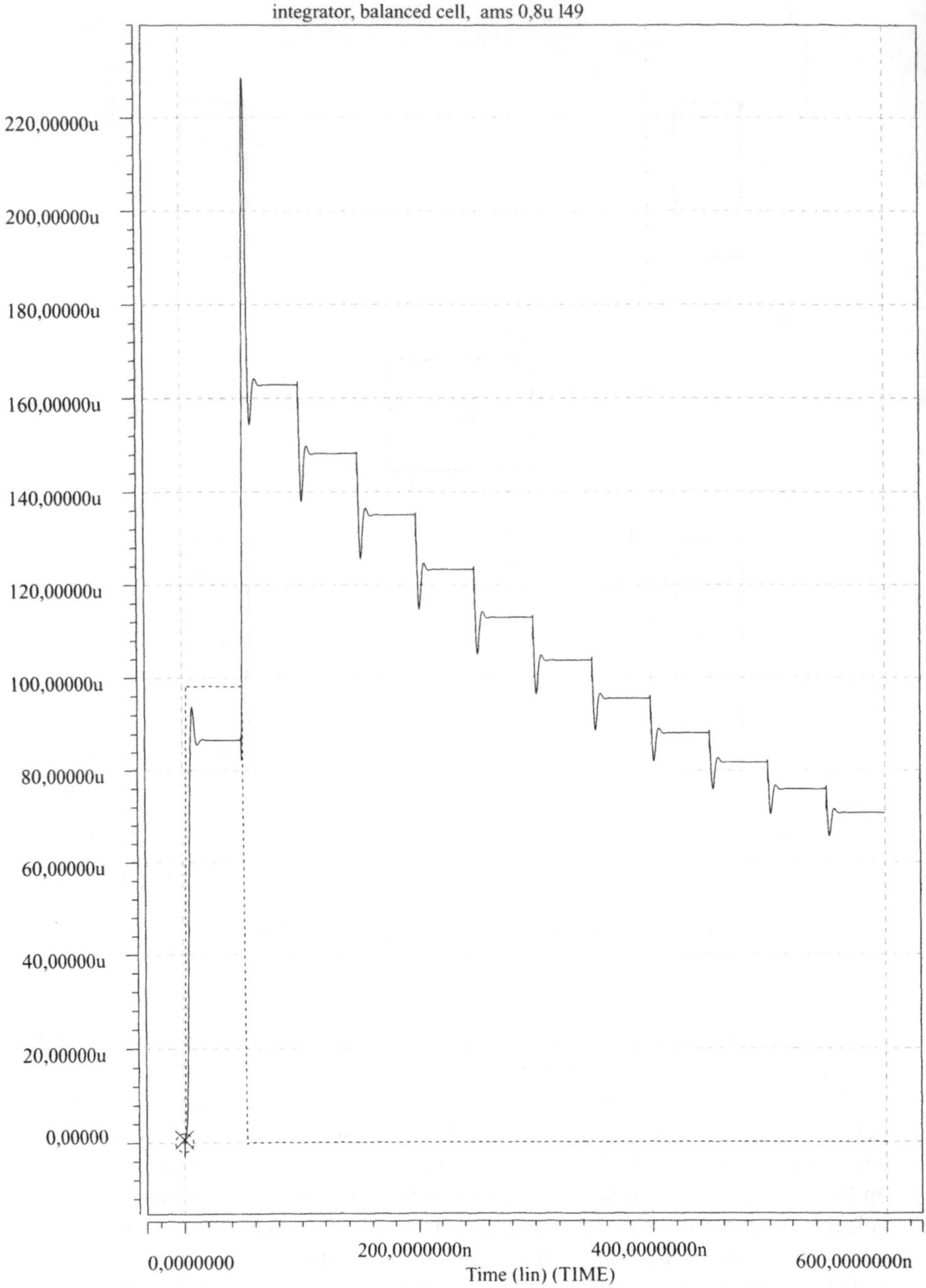

Figure 5.9 Impulse response current signal of the bilinear integrator based on Euler structures.

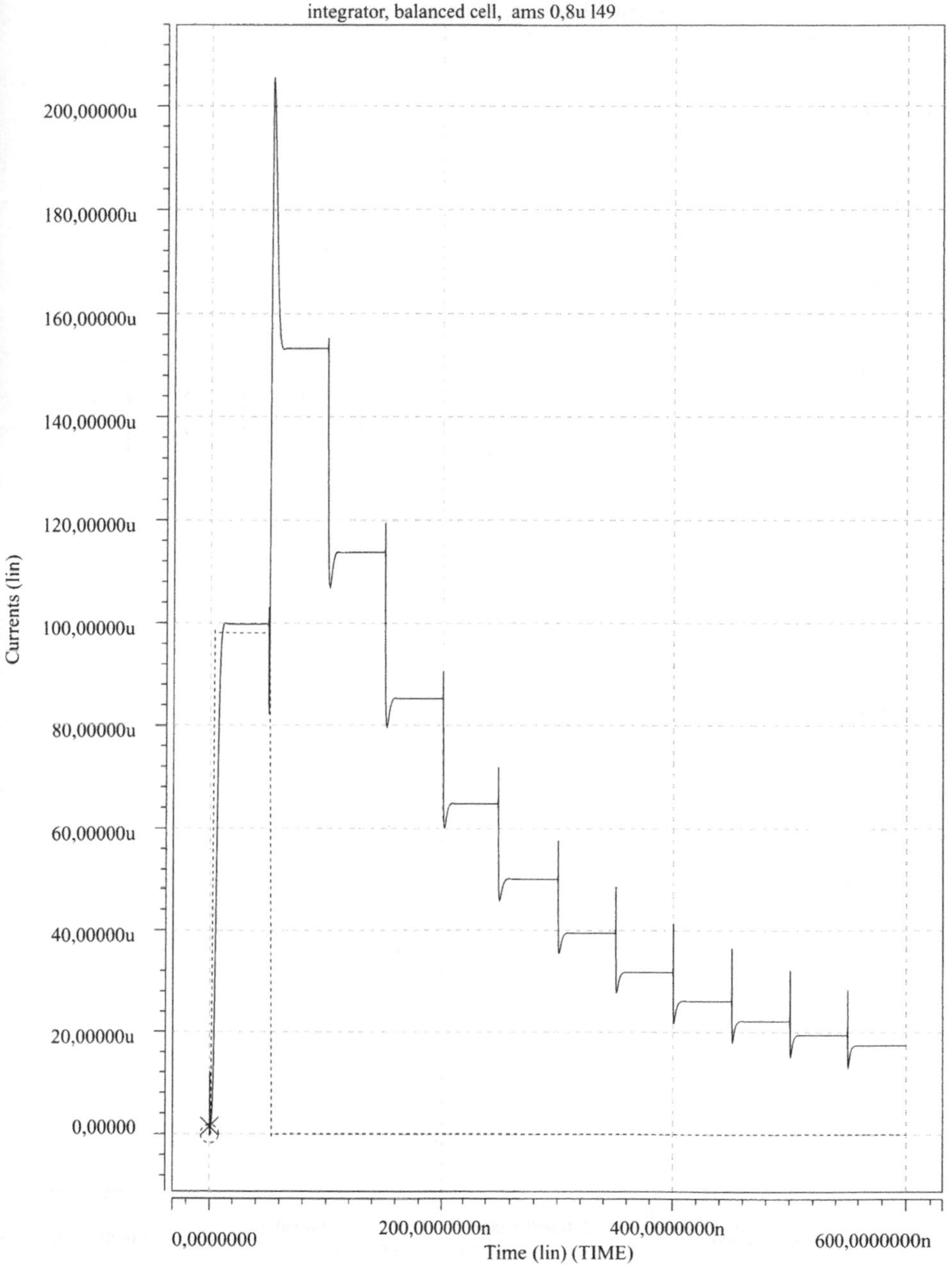

Figure 5.10 Impulse response of bilinear classic integrator without the gain error compensation.

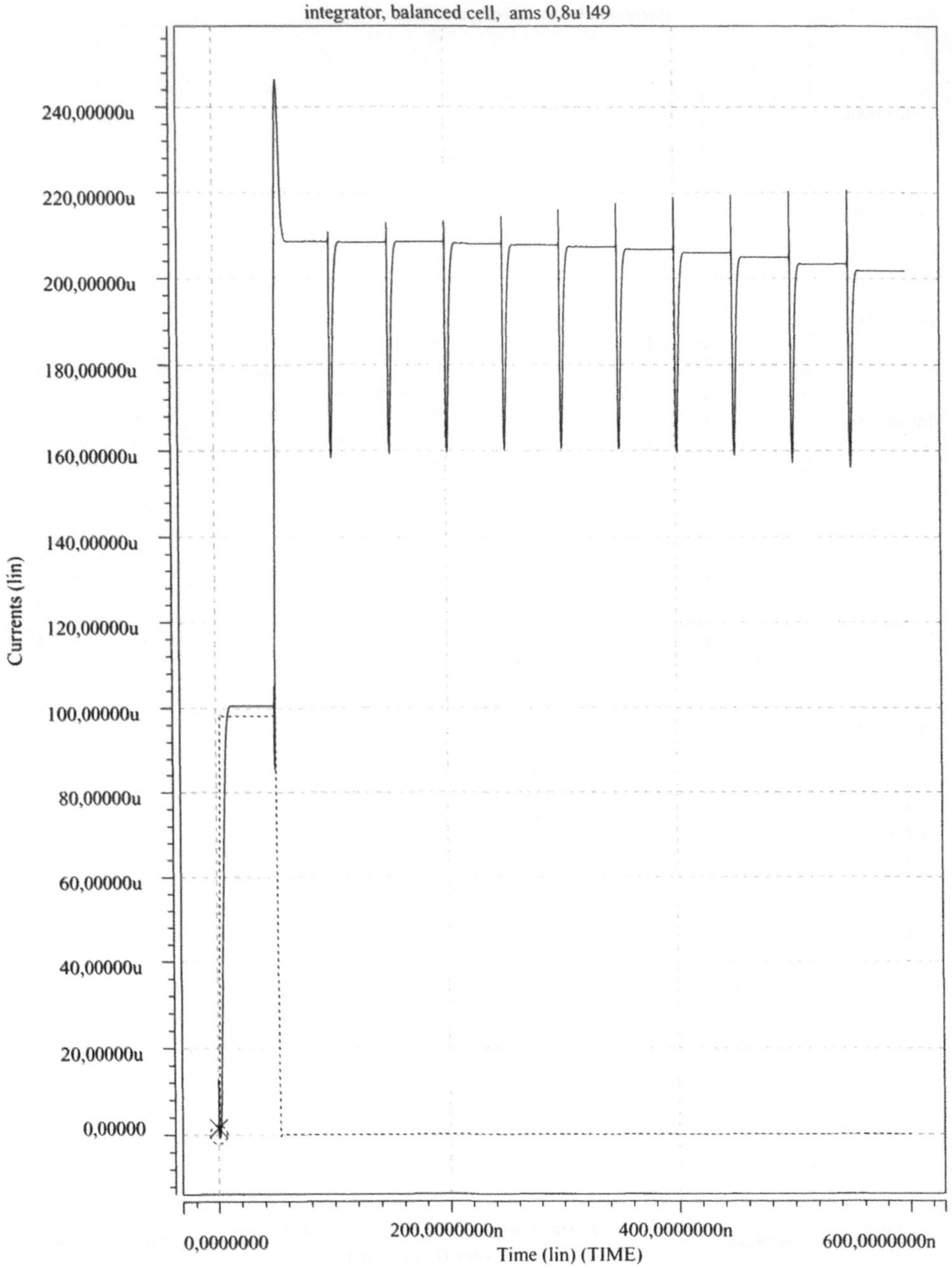

Figure 5.11 Impulse response of bilinear classic integrator with the gain error compensation.

5.4 SWITCHED CAPACITOR TECHNIQUE

In the previous section, we described SI circuits with discrete-time signals. Now we will present switched capacitor (SC) circuits. SC circuits also operate on discrete-time signals. However, unlike SI circuits, they work in the voltage mode and the basic active element is an operational amplifier. As was mentioned in Section 5.2, they can process signals as precisely as digital circuits. However, they need an additional, second poly layer in comparison to a standard digital CMOS process. Recently, a new type of SC circuit has been proposed [61], in which the gate-to-channel capacitance of MOS transistors is used for realizing all capacitors.

5.4.1 Integrators

The basic building blocks of switched capacitor circuits are integrators, similar to the switched current technique. SC integrators are composed of op-amps, capacitors, and switched capacitors. Two kinds of such integrators are shown in Figure 5.12.

Let us determine the transfer function of the integrator presented in Figure 5.12a when the switches are controlled by the clock phases shown without the parentheses. Let us assume that the jump of the input voltage V_{in} occurs at the beginning of the first phase and is held during both phases of the clock period. The inverting input of the op-amp has the ground potential (virtual ground). Hence, in phase Φ_2 the capacitance C_1, which was discharged in Φ_1, obtains the charge $\Delta q = C_1 v_{in}(n)$. The variable n in $v_{in}(n)$ means that signals are sampled during the second phase of each

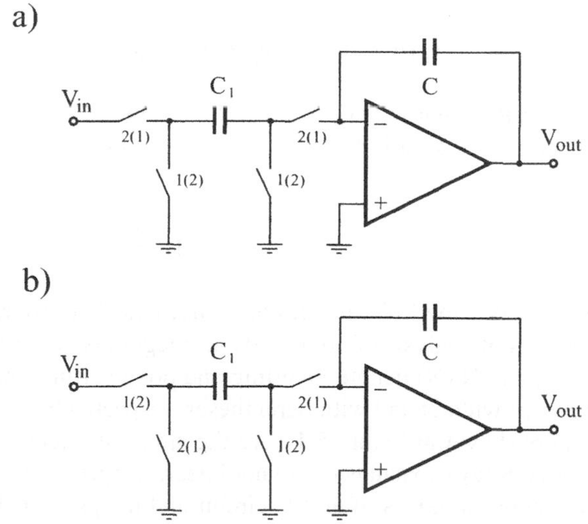

Figure 5.12 SC Euler integrators: inverting (a) and noninverting (b).

period. The same charge is injected into the feedback capacitance C in the inverting node of the op-amp, which is described by the equation

$$C_1 v_{\text{in}}(n) = -C[v_{\text{out}}(n) - v_{\text{out}}(n-1)] \tag{5.28}$$

The transfer function

$$H(z) = \frac{V_{\text{out}}(z)}{V_{\text{in}}(z)} = \frac{C_1}{C}\frac{-1}{1-z^{-1}} \tag{5.29}$$

shows that the SC circuit in Figure 5.12a is an inverting backward Euler integrator when the switches are controlled by the clock phases shown without the parentheses and when signals are sampled during the second phase of each clock period.

Assuming that signals are sampled during the first phase of the clock period, we obtain equation (5.28) in the form

$$C_1 v_{\text{in}}(n-1) = -C[v_{\text{out}}(n) - v_{\text{out}}(n-1)] \tag{5.30}$$

Hence,

$$H(z) = \frac{V_{\text{out}}(z)}{V_{\text{in}}(z)} = \frac{C_1}{C}\frac{-z^{-1}}{1-z^{-1}} \tag{5.31}$$

and the integrator realizes the forward Euler transform.

The analysis of the circuit shown in Figure 5.12b is similar. Choosing the clock phases shown in the parentheses, we can write the equation for the charge injected into the feedback capacitance in the form

$$-C_1 v_{\text{in}}(n-1) = -C[v_{\text{out}}(n) - v_{\text{out}}(n-1)] \tag{5.32}$$

assuming that both the input voltage jump and the sampling of signals occur in the second phase of each clock period. Hence, the transfer function

$$H(z) = \frac{V_{\text{out}}(z)}{V_{\text{in}}(z)} = \frac{C_1}{C}\frac{z^{-1}}{1-z^{-1}} \tag{5.33}$$

shows that the noninverting Euler integrator realizes the forward transform.

Complete results of analysis of both kinds of integrators are shown in Table 5.1. The symbols I (I) and N (N) denote inverting and noninverting integrators with the clock phases shown without and with parentheses, respectively.

The integrators shown in Figure 5.12 are the simplest integrators insensitive to stray capacitances. Stray capacitances are nonlinear, temperature-dependent capacitances occuring between nodes of an SC circuit and the ground. The stray capacitances consist of the capacitances of the top and bottom plates of capacitors in the SC circuit with respect to the ground and parasitic capacitances of transistors. In the

Table 5.1 Euler integrators

No.	Phase of input signal step	Kind of integrator	Phase of sampling	Euler transform
1	2	N	1 or 2	forward
2	2	(N)	1	backward
3	2	(N)	2	forward
4	1	N	1	forward
5	1	N	2	backward
6	1	(N)	1 or 2	forward
7	2	I	1 or 2	backward
8	2	(I)	1	backward
9	2	(I)	2	forward
10	1	I	1	forward
11	1	I	2	backward
12	1	(I)	1 or 2	backward

Euler integrators under consideration, we take into account four parasitic capacitances connected to the nodes of capacitances C_1 and C. As an example, let us analyze the stray capacitance between the input node of C_1 and the ground. This stray capacitance is periodically charged from the input source and then discharged to the ground and does not influence the charge in the signal path between the input and output of the integrator. Assuming an ideal op-amp, with virtual ground at the inverting input and with an ideal source at the output node, we can conclude that none of the remaining stray capacitances influence the operation of the circuit, either.

As in SI circuits, the clock feed-through effect is another important parasitic effect in SC integrators. This effect can be cancelled in the balanced structure of the integrator, as shown in Figure 5.13. The charge equation for the positive signal path (the clock phases that control the switches are shown without parentheses) can be written in the form

$$C_1\{[v_{in}(n)/2 - v - -v_{in}(n-1)/2 - v]\} = -C\{[v_+(n) - v - v_+(n-1) - v]\} \quad (5.34)$$

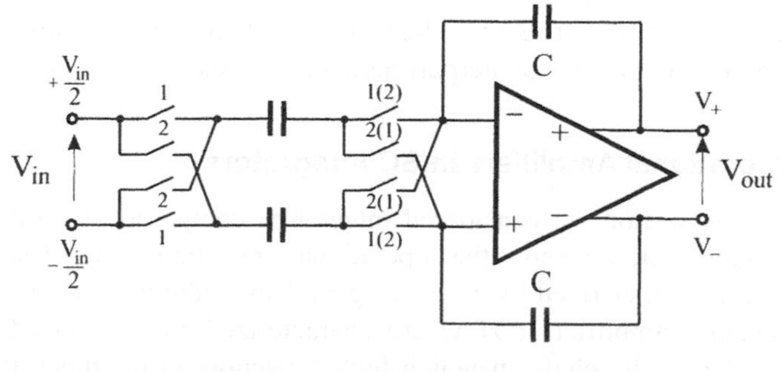

Figure 5.13 Bilinear balanced SC integrator.

and for the negative signal path in the form

$$C_1\{[-v_{\text{in}}(n)/2 - v - v_{\text{in}}(n-1)/2 - v]\} = -C\{[v_-(n) - v - v_-(n-1) - v]\} \quad (5.35)$$

Since

$$v_{\text{out}} = v_+ - v_- \quad (5.36)$$

after subtracting equation (5.35) from equation (5.34), we obtain the relation

$$C_1\{[v_{\text{in}}(n)] - [-v_{\text{in}}(n-1)]\} = -C\{[v_{\text{out}}(n) - v_{\text{out}}(n-1)]\} \quad (5.37)$$

and

$$H(z) = \frac{V_{\text{out}}(z)}{V_{\text{in}}(z)} = -\frac{C_1}{C}\,\frac{1+z^{-1}}{1-z^{-1}} \quad (5.38)$$

Similar analysis of the circuit presented in Figure 5.13, with clock phases shown in the parentheses, yields the noninverting bilinear integrator.

Let us note that in the above equations the signals are sampled in each half clock period. Hence, the sampling frequency is twice as great as the clock frequency. Relation (5.36) shows that the clock feed-through effect and crosstalk from digital signals are cancelled in the output signal.

However, introducing balanced structure in SC circuits means doubling the number of capacitors and, in consequence, enlarging the chip area. Moreover, converters are necessary to obtain balanced input signals and to realize the subtraction operation (5.36).

We can show that the SC integrators described by transfer functions (5.29), (5.33), and (5.38) are discrete counterpart networks of an analog integrator (5.17), realizing the backward Euler transformation, the forward Euler transformation, and the bilinear transformation. Let us note that the relation between conductance G_i of the input branch of an analog integrator and the switched capacitor C_1 is given by $C_1 = G_iT$ for Euler integrators and by $C_1 = G_iT/2$ for a bilinear integrator.

In many applications, we need an SC counterpart network of the damping integrator with the transfer function (5.23). In order to obtain damping integrators, the SC counterpart networks of resistors have to be included in the feedback branch of an SC integrator. These SC counterpart networks are shown in Figure 5.14.

5.4.2 Operational Amplifiers in SC Integrators

In the previous section we considered integrators composed of ideal operational amplifiers. However, we know that operational amplifiers realized in the CMOS technology are characterized by a limited gain–bandwidth product and operational transconductance amplifiers (OTA) are characterized by a limited transconductance. Considering the phase in which both capacitors in the input and feedback branches are connected to the amplifiers in the integrator, we can analyze the cir-

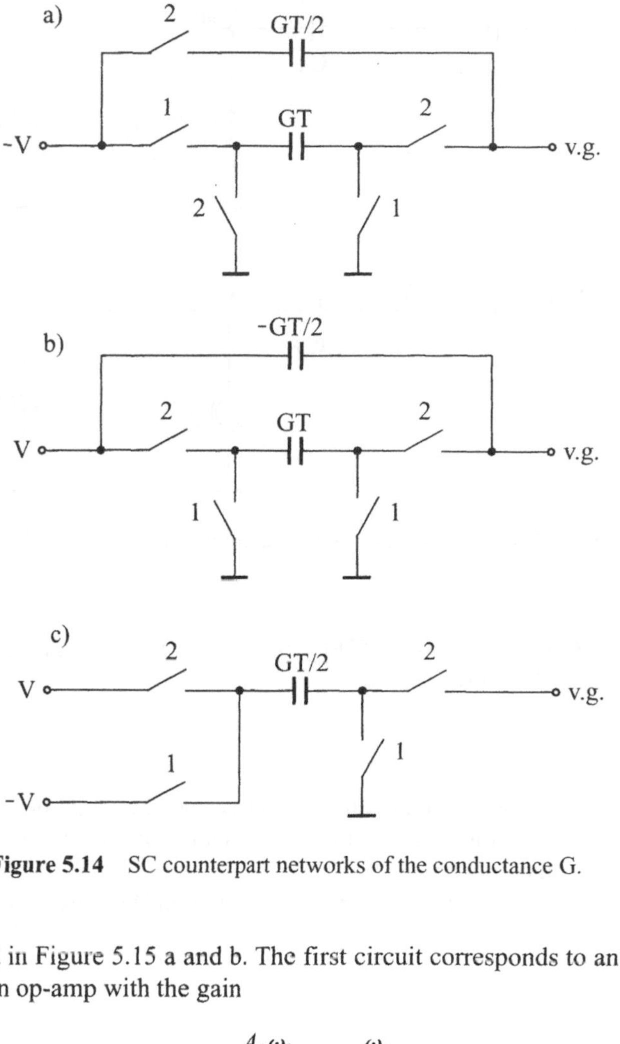

Figure 5.14 SC counterpart networks of the conductance G.

cuits presented in Figure 5.15 a and b. The first circuit corresponds to an integrator composed of an op-amp with the gain

$$A(s) = \frac{A_0\omega_0}{s + \omega_0} = \frac{\omega_1}{s + \omega_0} \tag{5.39}$$

where the gain–bandwidth product is denoted by ω_1 [49], and the second circuit corresponds to an integrator composed of an OTA with the transconductance g_m. C_i and C_f are input and feedback branch capacitances with the values depending on the clock phase. C_o denotes the capacitor at the output of the op-amp.

The voltage transfer function of the circuit in Figure 5.15 a is described by the formula

$$H(s) = \frac{V_{\text{out}}}{V_{\text{in}}} = -\frac{C_i}{C_f}\frac{1}{1 + \dfrac{\omega_0}{\omega_1}\dfrac{C_i + C_f}{C_f} + \dfrac{s}{\omega_1}\dfrac{C_i + C_f}{C_f}} = H_{\text{id}}\frac{1}{1 + \dfrac{\omega_0}{\omega_1}\beta + \dfrac{s}{\omega_1}\beta} \tag{5.40}$$

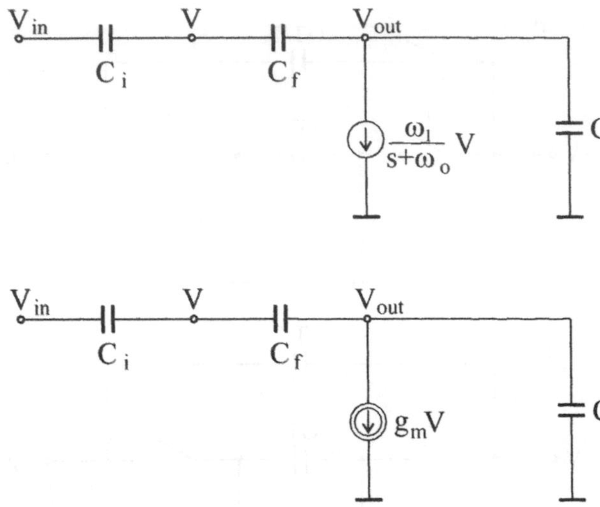

Figure 5.15 Models of integrators composed of an op-amp (a) and an OTA (b).

where $H_{id} = -C_i/C_f$ and $\beta = (C_f + C_i)/C_f$. Since in (5.39) $\omega_1 \gg \omega_0$, $A_0 \gg 1$, and $\beta \approx 1$, the formula (5.40) shows that for

$$\omega \ll \omega_1 \tag{5.41}$$

an op-amp with the gain–bandwidth product ω_1 has the properties of an ideal op-amp with the transfer function H_{id}. Let us note that C_o does not affect the behaviour of SC circuits containing op-amps.

For the circuit from Figure 5.15 b, composed of an OTA, we obtain

$$H(s) = \frac{V_{out}}{V_{in}} = \frac{C_i(sC_f - g_m)}{s(C_iC_f + C_iC_o + C_fC_o) + g_mC_f} = H_{id}\frac{1 - s\dfrac{C_f}{g_m}}{1 + s\dfrac{C_{eff}}{g_m}} \tag{5.42}$$

where $C_{eff} = C_i + C_o + C_iC_o/C_f$. Hence, for the transconductance amplifier the condition for ideal operation has the form

$$\omega \ll \omega_p = \frac{C_{eff}}{g_m} \tag{5.43}$$

For the parameter values $g_m = 100\mu A/V$ and $C_{eff} = 10pF$, typical for submicron technologies, we have $\omega_p = 10^9$, which is much greater than ω_1 of an op-amp realized in the same technology.

In SC circuits containing OTAs, we are able to reduce the settling time in switching states. Settling time is a very important parameter because it determines the maximum speed of circuit operation. In order to increase operating frequency, it is necessary to reduce the settling time. To achieve that, we use large-transconductance amplifiers and small-value capacitors. The settling behavior of a single amplifier, loaded by capacitors C_o, C_i, and C_f, is presented in Figure 5.16 [18, 58]. Figure 5.16 shows settling time t_s of the voltage at the output of the amplifier as a function of C_{eff}/g_m. Similar characteristics are also observed in the case of switched-current cells [1]. We see from this characteristic that there exist two regions of capacitance values. In the first, acceptable area, where $C_{eff} > C_1$, the behavior of the amplifier is typical: the settling time, which is proportional to C_{eff}/g_m, decreases with decreasing C_{eff}. For a given transconductance g_m, the limit capacitance C_1 denotes the border of the acceptable region. In the second region, where $C_{eff} < C_1$, and where loading capacitors are comparable to parasitic ones, we cannot consider the circuit as a first-order network. The poles of the amplifier dominate and cause oscillations. Hence, the settling time rises. Similar behavior can be expected in filters in which two or more operational amplifiers are coupled together in switching states. However, in this case we ought to take into consideration settling times at the outputs of all amplifiers. The settling behavior depends not only on the loading capacitor of the amplifier but also on the output capacitors of other amplifiers. It is impractical to find the characteristics of such a filter as functions of many variables and to use them for the design of the filter. Hence, in reference [22] we presented a more general approach with the use of an optimization process.

The main goal in high-frequency filter design is to achieve a minimum value of settling time. This goal is realized by adding grounded capacitors at outputs of amplifiers and by optimizing transconductances of amplifiers so that the settling time is minimized [18]. These additional capacitors, like parasitic capacitors, do not alter the transfer function of the filter. In [18] the authors present an optimization method that introduces some simplifications. The most important simplification concerns

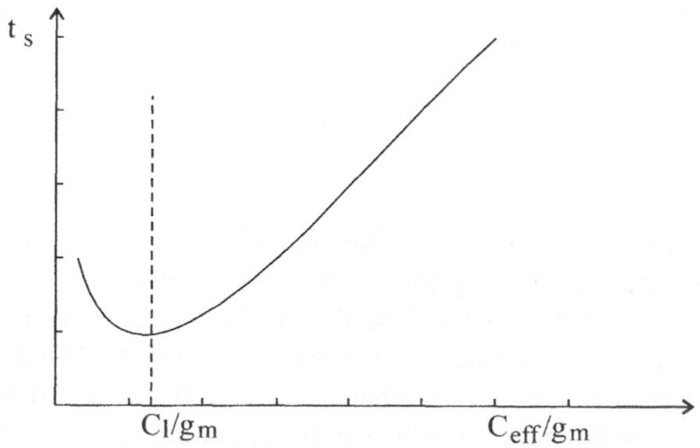

Figure 5.16 Transconductance amplifier settling time.

amplifiers. It is assumed that if in a certain switching state some of the amplifiers are coupled together, settling characteristics of each amplifier are investigated one by one, with the assumption that the remaining amplifiers are ideal. The idea proposed in [18], in the case of a capacitatively coupled multiamplifier network, is to analyze such a network using artificial intelligence techniques.

The minimization of the settling time, based on SPICE analysis of the SC network in switching states, is a time-consuming process. Hence, in order to improve the design technique, the solution of the problem presented in [22] is based on a topological method of analysis and can be used for an SC circuit containing an arbitrary number of operational amplifiers coupled together in each switching state and modeled as ideal transconductances. In Chapter 8, we present topological methods of analysis of networks composed of transconductors and capacitors. Computer tools based on these methods are also described in this chapter.

The objective function for settling time minimization is constructed in a symbolic form on the basis of a topological method of analysis. This function depends on optimization parameters that are capacitances of grounded capacitors and transconductances of VCTs. The computer tools based on topological methods are illustrated with an example of a fifth-order SC ladder filter design in [22]. SPICE simulations are also presented in [22] to show the efficiency of chosen tools.

5.5 LOW-POWER AND LOW-VOLTAGE TECHNIQUES

The average power consumption P of a digital circuit implemented in the CMOS technology is given by the relation

$$P = \alpha C_L V_{DD}^2 f \tag{5.44}$$

where V_{DD} is the supply voltage, f is the system clock frequency, and $\alpha\, C_L$ is the effective switched capacitance determined by the load capacitance C_L and the switching activity α [7]. The above formula shows that the power is the square of the supply voltage. Hence, the reduction of power supply voltage is the main goal for low-power design. However, the circuit delay τ described by the relation

$$\tau = \frac{2C_L V_{DD}}{\beta_n (V_{DD} - V_T)^2} \tag{5.45}$$

where β_n is given by (2.2), increases when the voltage decreases. In order to illustrate estimation of power consumption on the basis of these relations, let us consider an example. Assume that two independent tasks A and B need to be executed with the execution time $t_e = 4$ each, at the supply voltage 3 V. The deadline for task A, which arrives before B, is $t_{dA} = 5$, whereas for task B $t_{dB} = 10$. If we consider the application of a stand-by period for the system operating at 3 V, both tasks can be executed in the interval [0, 8] and the system can be in the stand-by mode in the interval [8, 10], giving a 20% power reduction.

The execution time $t_e(V_{DD})$ for the reduced voltage V_{DD} can be described by the formula

$$\frac{t_e(V_{DD})}{t_e(3)} = \frac{V_{DD}(3-V_T)^2}{3(V_{DD}-V_T)^2} \tag{5.46}$$

obtained from (5.45). For $t_e(V_{DD}) = 5$ and $t_e(3) = 4$ we see that the supply voltage can be reduced to $V_{DD} = 2.6$ V, assuming the threshold voltage $V_T = 0.6$ V. In this case task A is executed in the interval [0, 5] and B in [5, 10]. Since the frequency f is inversely proportional to the delay τ, equation (5.44) implies the relation

$$\frac{P(V_{DD})}{P(3)} = \frac{V_{DD}(V_{DD}-V_T)^2}{3(3-V_T)^2} \tag{5.47}$$

giving $P(V_{DD})/P(3) = 0.6$, which is equivalent to 40% power reduction.

The scaled CMOS process, in which supply voltage is reduced, ensures the decrease of power consumption in digital circuits. However, the decreasing of the threshold voltage V_T in this process increases the subthreshold current and power dissipation in the stand-by mode [42]. In order to overcome this difficulty, multi-threshold-voltage circuits, which need a nonstandard CMOS technology, are proposed [32]. The pipelining approach, implemented in the standard digital CMOS technology, can be used to achieve the same speed of operation as for scaled-down threshold voltage, at the cost of increased latency [32]. Another reason to keep the transistor threshold voltage V_T almost constant is to achieve a sufficient noise margin in dynamic digital circuits, [53].

Due to the current mode operation of SI circuits, the supply voltage reduction does not influence significantly their behavior. Equations (5.10) and (5.12) show that the circuits based on the complementary pair and used in the current mode are suitable for low-voltage operation. The operation voltage V_o ought to be greater than the absolute value of the threshold voltages, $V_o > V_{Tn}$, $V_o > -V_{Tp}$, to keep the transistors in saturation regions and to obtain positive transconductances. It means that the reduction of power supply voltage $V_{DD} - V_{SS} = 2V_o > V_{Tn} - V_{Tp}$ to 3 V is possible for typical values of threshold voltages lower than 1 V.

As an example of SI circuit operated at low-voltage supply, let us consider the basic memory cell in Figure 5.3c. In the first clock phase, in which gates and drains of transistors are connected to each other, the voltages between drains and sources are greater than the saturation voltages described by relations (2.4) and (2.10). Denoting the voltages as in Figure 5.4, we can write these inequalities in the form:

$$V > V - V_{Tn}, \quad -V > -V + V_{Tp} \tag{5.48}$$

for nMOS and pMOS transistors, respectively. Hence, both transistors in this phase are in the saturation region. In the second clock phase, relations (5.48) take the form

$$V' > V - V_{Tn}, \quad -V' > -V + V_{Tp} \tag{5.49}$$

where V is the voltage at the gates and V' is the voltage at the drains of those transistors whose gates and drains are disconnected in this phase. To keep the transistors in the saturation region, the condition obtained from (5.49)

$$V_{Tp} < \Delta V < V_{Tn}, \qquad \Delta V = V - V' \tag{5.50}$$

ought to be fulfilled. The voltage fluctuation ΔV depends on the output current of the memory cell considered and on the load being the input stage of the next cell. However, for typical dimensions of transistor channels and for the current range up to several hundreds of μA, relation (5.50) is fulfilled. Concluding, we can state that both digital circuits and SI circuits are able to operate at low-voltage supply.

This compatibility does not exist for digital circuits and SC circuits. The lack of compatibility is caused by the use of operational and transconductance amplifiers as basic elements. Modification in design is necessary when they are fabricated on the same chip in mixed-signal, low-voltage circuits. As an example, let us consider the differential stage presented in Figure 2.25, which is an input stage of each amplifier. Let us assume that the voltages V_1 and V_2 are equal to the common mode input voltage. For $V_1 = V_2$, the drain currents of the transistors M_{n1} and M_{n2} are equal to $I_B/2$, and from (2.82) and (2.83) we obtain

$$V_{GS1} = V_{GS2} = \sqrt{\frac{I_B}{\beta_n}} + V_{Tn} \tag{5.51}$$

The saturation voltage of the transistor M_{nB} can be obtained from (2.4) as

$$V_{DSB,\text{sat}} = V_{GSB} - V_{Tn} = \sqrt{\frac{2I_B}{\beta_n}} \tag{5.52}$$

Hence, the input common mode range, *CMR*, defining the range of common mode input voltage that ensures proper operation of the cell, is given by

$$CMR = V_{DD} - \left(\sqrt{\frac{2I_B}{\beta_n}} + \sqrt{\frac{I_B}{\beta_n}} + V_{Tn}\right) \tag{5.53}$$

where V_{DD} is the voltage delivered to the drains of transistors M_{n1} and M_{n2}. The component in the parentheses can be equal to 1.5 V or more. Hence, for $V_{DD} = 3$ V and *CMR* is only 50% of the supply voltage. The pMOS differential stage complements the nMOS one in the range of the common mode input voltage. When *CMR* is equal to the power supply voltage, $CMR = V_{DD}$, the cell has a rail-to-rail voltage range. This case can be achieved for a differential input stage composed of nMOS and pMOS stages, as shown in Figure 5.17. The total transconductance of such a cell is equal to the sum of the transconductances of both differential stages. In order to obtain the constant value of the total transconductance, additional circuits must be added. Such circuits and rail-to-rail output stages of operational amplifiers can be found in the literature [50].

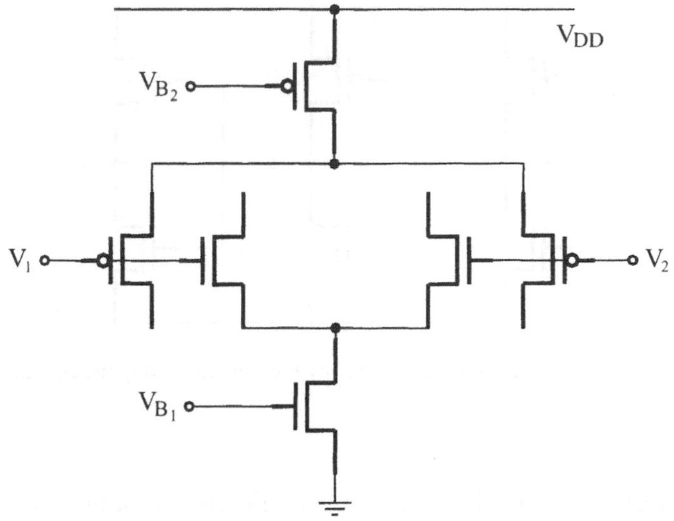

Figure 5.17 Rail-to-rail differential stage.

5.6 PROBLEMS

1. Calculate the power $p(t) = v(t)i(t)$ delivered to the resistance R and energy $E(t)$ delivered to reactance elements: inductance L and capacitance C. Prove that R, L, and C are passive elements.
2. For the following equation of a two-port network

$$\begin{bmatrix} U_1 \\ I_1 \end{bmatrix} = \begin{bmatrix} a & b \\ c & d \end{bmatrix} \begin{bmatrix} U_2 \\ I_2' \end{bmatrix} \tag{5.54}$$

described with the use of the chain matrix composed of the elements a, b, c, and d, calculate the input impedance Z_{in} of this two-port network loaded by the impedance Z_L. Prove that a converter has the matrix elements $b = 0$, $c = 0$, whereas an inverter has $a = 0$, $d = 0$. What are the values of the nonzero elements and the conversion (inversion) coefficient k_c, (k_i) of an ideal transformer (gyrator)?

3. Prove that the chain connection of CVT and VCT gives CCT, and that the chain connection of VCT and CVT gives VVT.
4. On the basis of equation (5.10) and assuming that $\beta_n = \beta_p$, prove that the relation between voltages V_1 and V_2 in the circuit in Figure 5.18 is given by $V_2 = -V_1 + (V_{Tn} + V_{Tp})$.
5. Calculate the transfer function (5.22) of the integrator shown in Figure 5.8.
6. Show that the transfer function (5.17) is equivalent to transfer functions (5.29), (5.33), and (5.38) for transformations (1.27), (1.26), and (1.28), re-

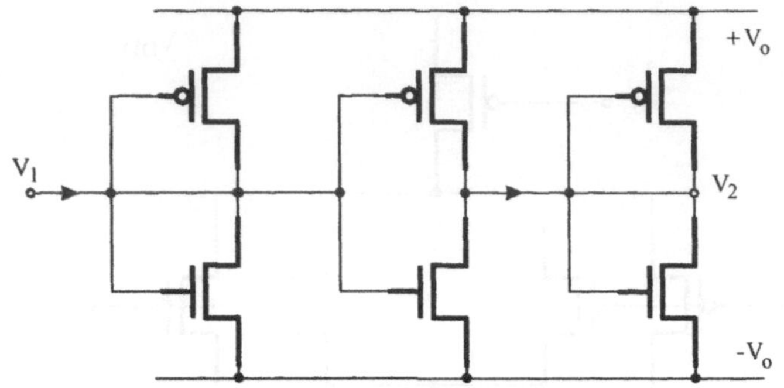

Figure 5.18 Circuit for the calculation of energy E dissipated during the charging/discharging of the load capacitance C_L.

spectively. What are the relations between G and C_1 for these kinds of integrators?

7. Prove that the switched-capacitor circuits shown in Figure 5.14 are bilinear counterpart networks of the conductance G.
8. Using the results obtained in problem 1 and assuming that the capacitance C_L, shown in Figure 5.19, is charged to the voltage V_{DD} during the first half of the

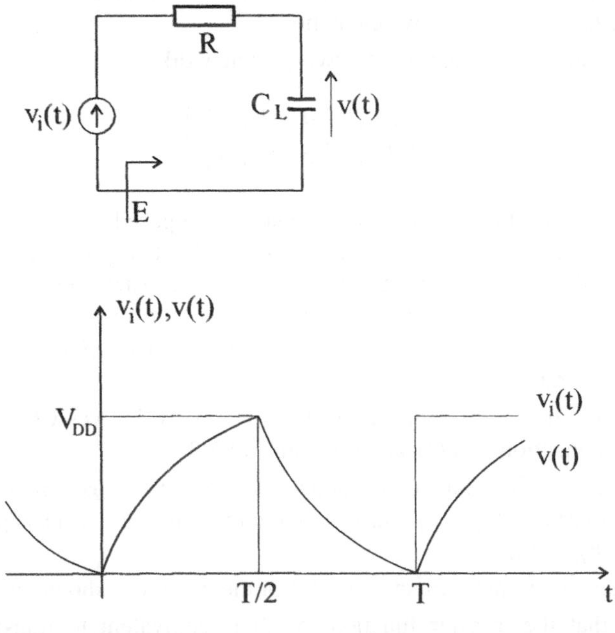

Figure 5.19 Example for the calculation of transmission error in cascaded cells.

clock period $T/2$ and completely discharged in the second half of the clock period, calculate the power dissipation given by (5.44) for $\alpha = 1$.

9. Using formulae (2.5) for the drain current of a MOS transistor in the saturated mode and for the charge $Q = C_L \cdot V_{DD}$ on the capacitance C_L, estimate the time needed for the charging process given by (5.45).

Part II

Methods and Tools for Mixed Signal Circuit Design

6

One-Dimensional Signal Processing

This chapter describes the low-sensitivity strategy in the designing of one-dimensional filters implemented in switched-current, switched-capacitor, and OTA-C techniques. The same strategy is used in digital filter design. We also present analog-to-digital converters based on delta-sigma modulation, which can be implemented in switched-current and switched-capacitor techniques.

6.1 CONVERTERS AND DELTA-SIGMA MODULATORS

Analog-to-digital (A/D) and digital-to-analog (D/A) converters are key components in many signal processing systems. An example of such a system was shown in Figure 1.1. The highest conversion rate of all converter architectures is provided by the flash converter [48]. A block diagram of the flash converter is shown in Figure 6.1. The converter is composed of $2^n - 1$ comparators, which are 1-bit A/D converters. The 1-bit converters operate in parallel connection and are followed by a digital circuit that produces an n-bit signal from the digital signal obtained in the thermometer code at the comparator outputs. The signal reference block gives $2^n - 1$ reference signals with the step $\Delta x_{\text{ref}} = x_{\text{ref}}/2^n$. Due to exponential dependence of the number of comparators on the number n of bits and due to analog component matching, no more than 10-bit resolution is typically obtained. This disadvantage does not occur in converters based on delta-sigma modulation ($\Delta\Sigma$ modulation) [54].

A delta-sigma modulator is shown in Figure 6.2. It consists of a summer, an integrator, a comparator, and a digital-to-analog converter. At its output [$y_i = sign(v_i)$], the comparator has only two quantization levels: "0" for $y_i \leq 0$ and "1" for $y_i = 1$. We will denote the quantization levels at the output of the comparator by $y_i \doteq 0$ or

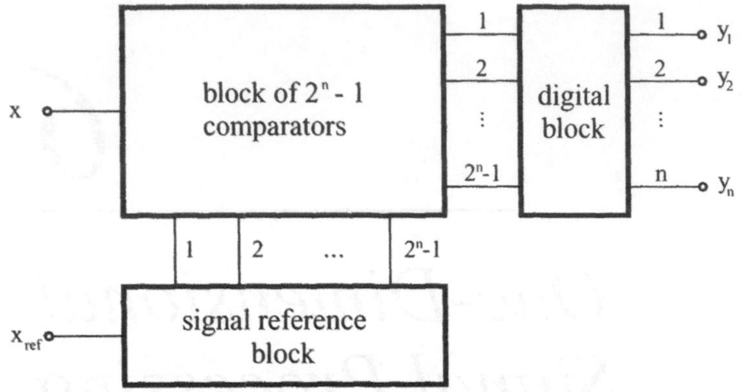

Figure 6.1 Block diagram of a flash converter.

$y_i \doteq 1$. Hence, the D/A converter is a simple 1-bit input element, described by the relation

$$f_i = \begin{cases} x_{\max}, \text{ for } y_i \doteq 1 \\ x_{\min}, \text{ for } y_i \doteq 0 \end{cases} \tag{6.1}$$

where $x_{\min}$ and $x_{\max}$ are minimum and maximum values of the sampled and held input signal $x_{\min} \leq x_k < x_{\max}$. The modulator is called the first-order modulator if it contains one integrator and one feedback loop. Second-order modulators contain two integrators in two feedback loops. Modulators with three and more feedback loops are not used because of stability problems. L-th order modulators, where $L > 2$, are composed of first- and second-order modulators. In contrast to flash converters, the $\Delta\Sigma$ modulators do not need high-precision analog components.

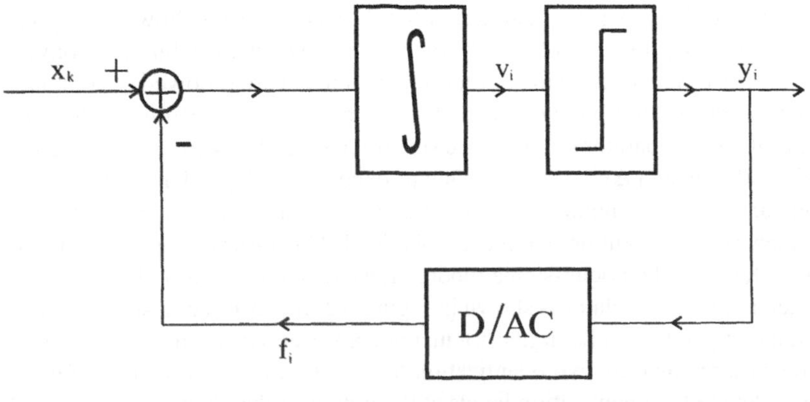

Figure 6.2 First-order delta-sigma modulator.

As we know from the first part of the book, all cells of a ΔΣ modulator can be easily implemented using SC or SI techniques. Let us explain the operation of the ΔΣ modulator, assuming that the integrator is a forward Euler integrator described by the transfer function (5.16). The signal v_i at the output of this integrator can be described by the equation

$$v_i = x_k - f_{i-1} + v_{i-1} \tag{6.2}$$

Assuming that $v_{-1} = 0$, $y_{-1} = 0$, and $f_{-1} = x_{min}$, we can calculate v_i, where $i = 0, \ldots N - 1$, for a given x_k. The number $N = 2^n$, for n-bit resolution, is called the oversampling ratio.

As an example, let us consider the nonnegative sampled and held signal x_k depicted in Figure 6.3. We assume that $x_{min} = 0$, $x_{max} = 8$, and $N = 8$. For a selected value of x_k, the signals v_i, y_i, and f_i, where $i = 0, \ldots 7$, can be calculated at the out-

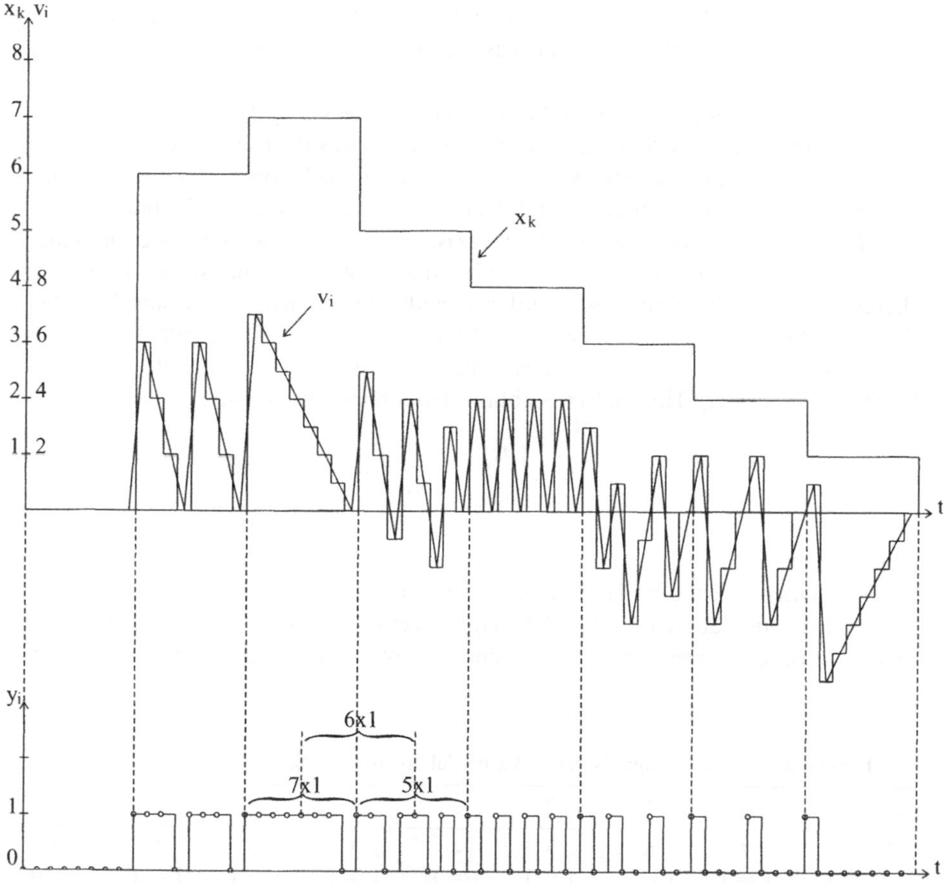

Figure 6.3 Example of signals in the first-order delta-sigma modulator.

Table 6.1 Values of signals in the ΔΣ modulator for $m = 8$, $x_{min} = 0$, and $x_{max} = 8$

x_k	i	0	1	2	3	4	5	6	7
3	v_i	3	−2	1	−4	−1	2	−3	0
	y_i	$1 \doteq 1$	$-1 \doteq 0$	$1 \doteq 1$	$-1 \doteq 0$	$-1 \doteq 0$	$1 \doteq 1$	$-1 \doteq 0$	$0 \doteq 0$
	f_i	8	0	8	0	0	8	0	0
7	v_i	7	6	5	4	3	2	1	0
	y_i	$1 \doteq 1$	$1 \doteq 1$	$1 \doteq 1$	$1 \doteq 1$	$1 \doteq 1$	$1 \doteq 1$	$1 \doteq 1$	$0 \doteq 0$
	f_i	8	8	8	8	8	8	8	0

puts of the itegrator, the comparator, and the converter, respectively. The results of such calculations for $x_k = 3$ and $x_k = 7$ are presented in Table 6.1. We observe that the number of "1"s in the sampling period is equal to the average value of the input signal.

The signal x_k can have positive and negative values. In order to illustrate this case, we assume that $x_{min} = -4$, $x_{max} = 4$, and $N = 8$. The results of calculations for $x_k = 3$ are presented in Table 6.2. In this case, the signal $x_k = 3$ has the same code as $x_k = 7$ in the previous example.

The resolution of an L-th order ΔΣ modulator increases by $L + 1/2$ bits when the oversampling ratio N is doubled. This dependence limits the use of such converters to low-frequency applications. A similar increase of resolution is obtained for converters realized in the parallel ΔΣ architecture. Each additional parallel branch with an L-th order modulator gives an L bit increase of resolution, with the oversampling ratio unchanged [34]. The architecture of such a converter, composed of N parallel channels, is given in Figure 6.4. In order to modulate and demodulate signals in the channels, the coefficients $u_{n,k}$ and $u_{n,k-1}$, for $n = 0, \ldots, N-1$, are calculated from the equation $u_{n,k} = m_{n,k \bmod N}$, where $m_{n,k}$ and $n, k = 0, \ldots, N-1$, are elements of a Hadamard matrix H_i. This matrix is obtained in a recurrent manner:

$$H_i = \begin{bmatrix} H_{i-1} & H_{i-1} \\ H_{i-1} & -H_{i-1} \end{bmatrix} \tag{6.3}$$

where $H_{-1} = [1]$.

The elements of the Hadamard matrix are ±1. Hence, the operation of modulation and demodulation is reduced to sign inversion of signals for negative elements. In order to use elements of a chosen row as the modulation sequence in

Table 6.2 Values of signals in the ΔΣ modulator for $M = 8$, $x_{min} = -4$, and $x_{max} = 4$

x_k	i	0	1	2	3	4	5	6	7
3	v_i	7	6	5	4	3	2	1	0
	y_i	$1 \doteq 1$	$1 \doteq 1$	$1 \doteq 1$	$1 \doteq 1$	$1 \doteq 1$	$1 \doteq 1$	$1 \doteq 1$	$0 \doteq 0$
	f_i	4	4	4	4	4	4	4	−4

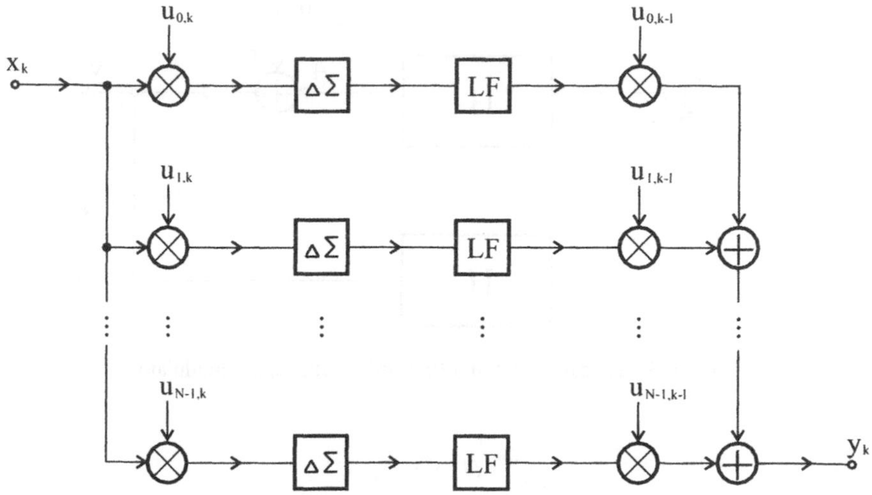

Figure 6.4 Parallel delta-sigma modulator architecture.

the corresponding channel, the number of rows must be equal to the number of channels. Hence, the equation $N = 2^{i+1}$ must be fulfilled. It is shown in [15] that the signal is simply delayed in the parallel architecture in Figure 6.4, while the noise of ΔΣ modulators is filtered in the low-pass filters *LF*. To illustrate this property, let us consider an $N = 2$ channel converter. The values of elements in the matrix

$$H_0 = \begin{bmatrix} 1 & 1 \\ 1 & -1 \end{bmatrix} \tag{6.4}$$

mean that the signals are always multiplied by 1 in the Hadamard modulator and demodulator of the first channel. The modulator in the second channel multiplies the signal alternately by 1 and –1, and the demodulator multiplies the signal alternately by –1 and 1, assuming the time-shifting of the demodulator $l = 1$. Hence, for second-order FIR filters described by the coefficients h_0, h_1, and h_2, we have $r_{0,k} = +1 \cdot h_0x_k + 1 \cdot h_1x_{k-1} + 1 \cdot h_2x_{k-2}$ at the output of *LF* in the first channel, whereas in the second channel we have $r_{1,k} = +1 \cdot h_0x_k - 1 \cdot h_1x_{k-1} + 1 \cdot h_2x_{k-2}$. As a result, at the output $y_k = +1 \cdot r_{0,k} - 1 \cdot r_{1,k} = 2 \cdot h_1x_{k-1}$ we obtain the delayed, one-bit representation of the input signal, scaled by the center coefficient of the FIR filter and multiplied by the number of channels. The other coefficients of the FIR filter can be chosen so that they ensure maximum attenuation of the noise.

Figure 6.5 presents a linear model of the first order delta-sigma modulator from Figure 6.2. The comparator is replaced by the summer, which introduces the noise n into the signal path; the integrator is replaced by the block with the transfer function H_1; and the feedback branch is represented by the block with the transfer function H_2. The response y of the circuit in Figure 6.5 can be calculated as

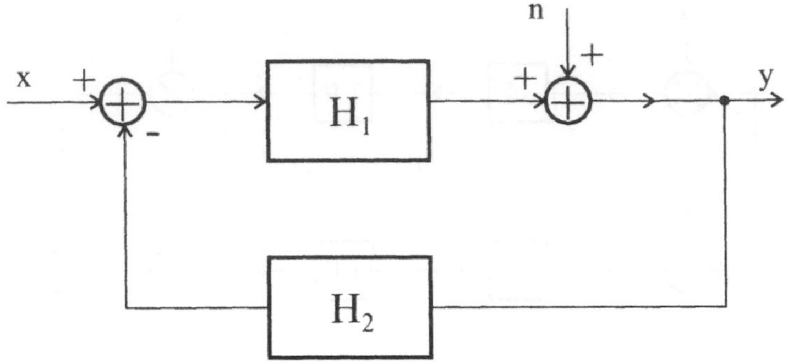

Figure 6.5 Linear model of a first-order delta-sigma modulator.

$$y = H \cdot x + H_n \cdot n \tag{6.5}$$

where

$$H = \frac{H_1}{1 + H_1 H_2}, \qquad H_n = \frac{1}{1 + H_1 H_2} \tag{6.6}$$

We normally set

$$H_1 = 1/s, \qquad H_2 = 1 \tag{6.7}$$

which corresponds to the modulator shown in Figure 6.2. In this setting the transfer functions (6.6) have the form

$$H = \frac{1}{s+1}, \qquad H_n = \frac{s}{s+1} \tag{6.8}$$

and the modulator transmits the signal x like a low-pass filter described by the transfer function H and attenuates the noise n like a high-pass filter described by the transfer function H_n. H_1, given by (6.7), denotes an integrator, which can be implemented in SC or SI techniques. A proper choice of transfer functions H_1 and H_2 leads to more complicated structures of modulators in comparison to that shown in Figure 6.2. The following sections show design methods of SC or SI filters that can be used as blocks described by transfer functions H_1 and H_2 in a delta-sigma modulator.

6.2 LOW-SENSITIVITY FILTER DESIGN STRATEGY

Filtering operations are very often used in signal processing systems. In this section, we will present the basics of the low-sensitive strategy of filter design. We will describe sensitivity properties of a lossless two-port network terminated by resistors, considered as a special case of a multiport network.

The voltage transfer function of a doubly loaded, lossless two-port network (Figure 6.6) can be written as a rational function in the form

$$H(s) = \frac{V_2(s)}{V_{in}(s)} = \frac{N(s)}{D(s)} \tag{6.9}$$

where $N(s)$, $D(s)$ are numerator and denominator polynomials. Assuming that the generator, the two-port network, and the load are all matched to each other, the maximum power delivered from the generator is

$$P_{max} = \frac{V_{in}}{4R_1} = \frac{V_{in}G_1}{4} \tag{6.10}$$

If the matching conditions are fulfilled, the power dissipated in the load of a lossless two-port network

$$P_2 = \frac{|V_2|^2}{R_2} = |V_2|^2 G_2 \tag{6.11}$$

is equal to P_{max} ($P_2 = P_{max}$). For nominal values of parameters, the two-port network can be matched to its terminations at a generator frequency ω_0. We assume that at least one such frequency exists. If matching conditions are not fulfilled because of frequency changes or circuit parameter variations, we have

$$\frac{P_{max}}{P_2} \geq 1 \tag{6.12}$$

or

$$P_{max} = P_2 + P_r \tag{6.13}$$

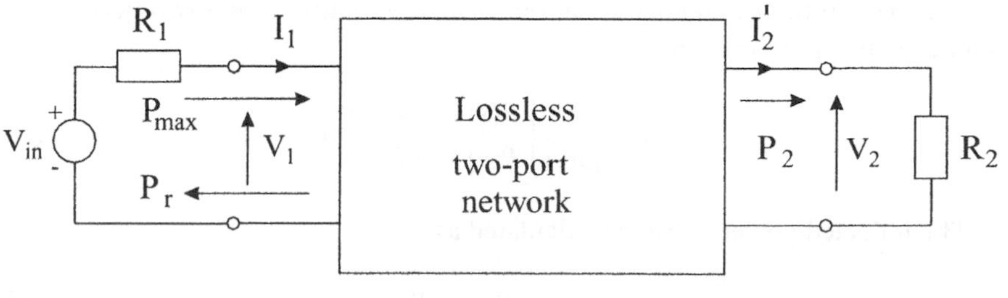

Figure 6.6 Lossless two-port network loaded by resistors.

where P_r is reflected power. P_r is nonzero when the matching conditions are not fulfilled. Introducing the transducer factor

$$H_t(s) = \frac{1}{2}\sqrt{\frac{G_1}{G_2}}\,\frac{1}{H(s)} = \frac{D(s)}{P(s)} \tag{6.14}$$

where $P(s) = 2\sqrt{\frac{G_2}{G_1}}N(s)$, we obtain

$$|H_t(j\omega)|^2 = \frac{P_{\max}}{P_2} \tag{6.15}$$

Similarly, we can introduce the characteristic function

$$K(s) = \frac{F(s)}{P(s)} \tag{6.16}$$

so that

$$|F(j\omega)|^2 = \frac{P_r}{P_2} \tag{6.17}$$

From relation (6.13) we obtain the Feldtkeller equation

$$|H_t(j\omega)|^2 = |K(j\omega)|^2 + 1 \tag{6.18}$$

which in the s domain becomes

$$H_t(s)H_t(-s) = K(s)K(-s) + 1 \tag{6.19}$$

With the use of the polynomials $D(s)$, $P(s)$, and $F(s)$, it is possible to rewrite equation (6.19) in the form

$$D(s)D(-s) = F(s)F(-s) + P(s)P(-s) \tag{6.20}$$

and to calculate, after factorization, the polynomial $F(s)$ on the basis on the given polynomials $N(s)$ and $D(s)$ of the voltage transfer function (6.9). On the basis of the polynomials in the Feldtkeller equation, scattering matrix S of a two-port network can be obtained in the form

$$S = \frac{1}{D(s)}\begin{bmatrix} F(s) & \pm P(-s) \\ P(s) & \mp F(-s) \end{bmatrix} \tag{6.21}$$

The reflected power P_r can be calculated as

$$P_r = P_{\max} - P_1 \tag{6.22}$$

where $P_{\max}$ is given by (6.10) and $P_1 = |I_1|^2 reZ_{\text{in}}$. Hence

$$P_r = P_{\max}\left(1 - \frac{4R_1}{|V_{\text{in}}|^2}|I_1|^2 reZ_{\text{in}}\right) = P_{\max}\left(1 - \frac{4R_1 reZ_{\text{in}}}{|R_1 + reZ_{\text{in}}|^2}\right) \tag{6.23}$$

and after simple calculations we have

$$\frac{P_r}{P_{\max}} = \frac{|K(j\omega)|^2}{|H(j\omega)|^2} = \left|\frac{R_1 - Z_{\text{in}}(j\omega)}{R_1 + Z_{\text{in}}(j\omega)}\right|^2 \tag{6.24}$$

The last equation gives the following relation in the s domain

$$\frac{K(s)}{H_t(s)} = \frac{F(s)}{D(s)} = \frac{R_1 - Z_{\text{in}}(s)}{R_1 + Z_{\text{in}}(s)} \tag{6.25}$$

or

$$Z_{\text{in}}(s) = R_1 \frac{D(s) - F(s)}{D(s) + F(s)} \tag{6.26}$$

Darlington proved that each positive real function can be realized as a driving-point impedance $Z_{\text{in}}(s)$ of a lossless two-port network terminated in a resistor. There are many synthesis methods for realization of a two-port network with the use of Z_{in}. In the realization procedure of such a two-port network on the basis of the voltage transfer function (6.9), the Feldtkeller equation plays the main role. Such an approach to synthesis, based on a lossless two-port network, is called the Darlington concept of circuit synthesis.

It follows from relations (6.12) and (6.15) that the magnitude of the transducer factor $H_t(j\omega)$ achieves the minimum value 1 for the frequencies in the passband in which the matching conditions are fulfilled. It means that the attenuation is equal to zero. If we assume $2\sqrt{G_2/G_1} = 1$, the magnitude of the transfer function $H(j\omega)$ achieves the maximum value 1. This case is presented in Figure 6.7. We see that the derivative of the transfer function magnitude $|H|$, and consequently the magnitude sensitivity

$$S_{x_i}^{|H|} = \frac{\partial \ln |H|}{\partial \ln x_i} = \frac{x_i}{|H|}\frac{\partial |H|}{\partial x_i} \tag{6.27}$$

is equal to zero with respect to parameters x_i. The smaller the ripple error in the passband, the smaller the magnitude sensitivity in this band is expected to be.

In the next section, we will describe synthesis methods of active circuits based on a lossless prototype network. With the use of these methods, we will design active filters that are easy to realize in the CMOS technology. We assume that the active filter obtained on the basis of the lossless prototype network inherits sensitivity properties of their prototype networks. The most popular prototype circuits are reactance lad-

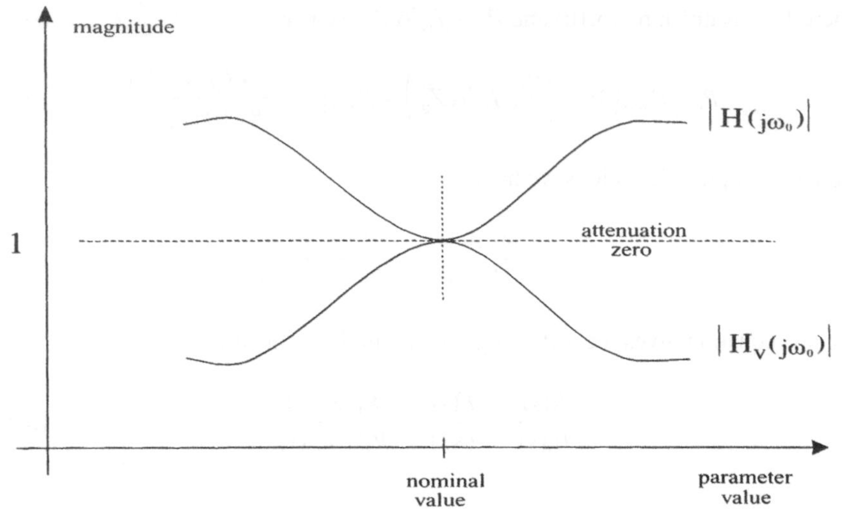

Figure 6.7 Transducer factor and voltage ratio as functions of a lossless circuit parameter.

der, two-port networks. Exact sensitivity formulae for this kind of two-port network were obtained and presented in [45, 46]. However, in this design strategy, only the transfer function with a purely even or odd numerator polynomial can be realized [55]. To synthesize filters and multiport networks like multiplexers, with zeros in the right-hand side of the complex plane, we need nonreciprocal prototype circuits. Hence, circuits composed of gyrators will be considered in the next sections.

6.3 OTA-C, SC, SI, AND DIGITAL FILTERS

It was shown in the previous section that a multiport lossless network terminated in resistors has good sensitivity properties. In this section, we will show an active filter design strategy in which filters inherit sensitivity properties from lossless prototype networks. A reactance ladder two-port network is very often used as a prototype network in the design of low-sensitivity digital, OTA-C, SC, and SI filters. It was shown in the first part of this book that such circuits, in contrast to reactance filters, are very easy to implement in the CMOS technology.

In order to explain the low-sensitivity active filter design strategy, we consider a third-order ladder filter shown in Figure 6.8a. The ladder filter is equivalent to the gyrator–capacitor network shown in Figure 6.8b obtained when the inductance L is replaced by two gyrators g and the capacitance $C = g^2L$. In the following subsection, we will show that the circuit in Figure 6.8c, composed of so called quasiideal gyrators, is also equivalent to the network in Figure 6.8a and that both gyrator–capacitor circuits can be effectively used in design of OTA-C, SC, SI, and digital filters.

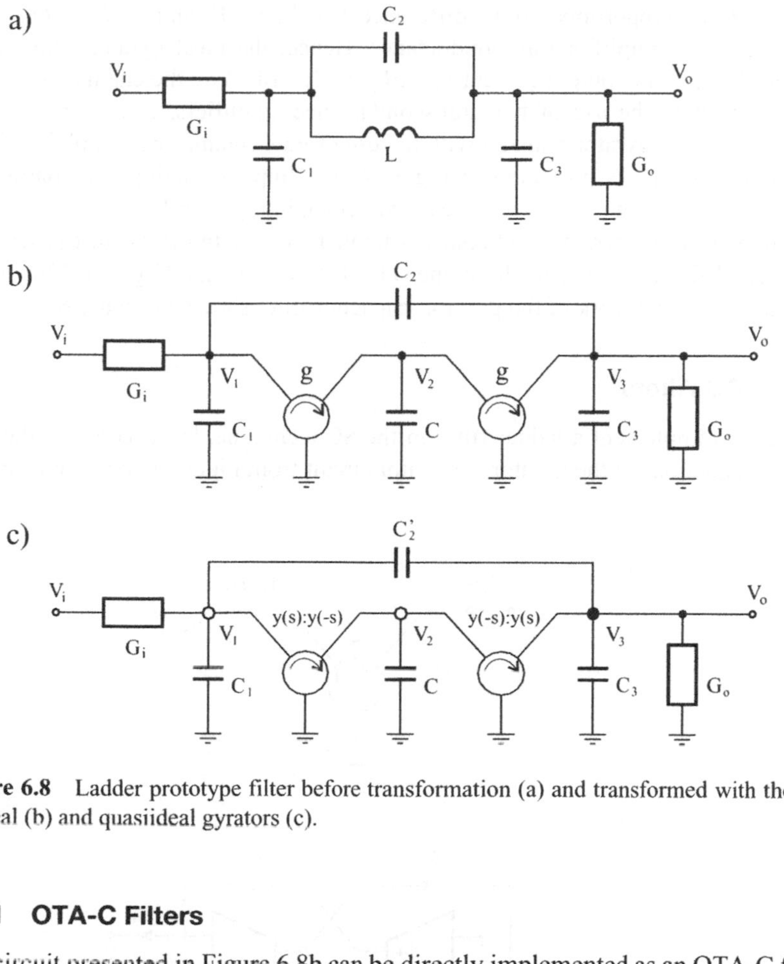

Figure 6.8 Ladder prototype filter before transformation (a) and transformed with the use of ideal (b) and quasiideal gyrators (c).

6.3.1 OTA-C Filters

The circuit presented in Figure 6.8b can be directly implemented as an OTA-C filter with the use of two transconductance amplifiers for each gyrator and two additional amplifiers for input and output conductances G_1 and G_2.

The fully differential transconductance amplifier presented in Figure 2.28, with the symbol shown in Figure 6.9, can be used in such an implementation. The output

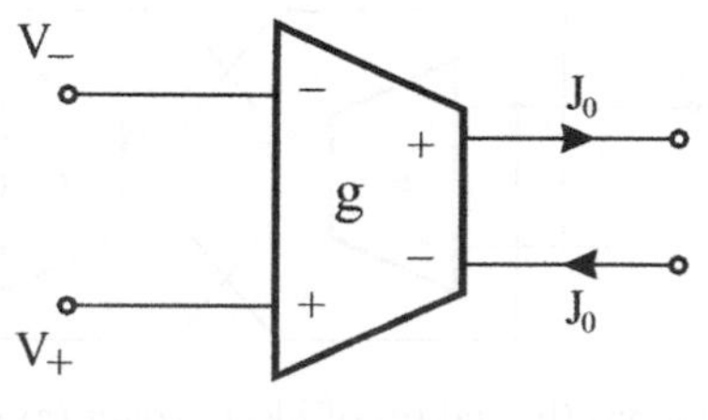

Figure 6.9 Symbol of a fully differential transconductance amplifier.

currents I_o are proportional to the difference of voltages V_+ and V_-, $I_o = g(V_+ - V_-)$, where g is the amplifier transconductance. Hence, the ideal gyrator whose input current $I_i = gV_o$ and output current $I_o = gV_i$ are described by the chain matrix (5.4), is realized with the use of two transconductance amplifiers, as shown in Figure 6.10b. The ideal gyrator realized with the use of transconductance amplifiers in the balanced structure is presented in Figure 6.10c. Implementations of floating and grounded resistors with conductances G are given in Figure 6.11.

On the basis of gyrators and resistors implemented with the use of transconductance amplifiers, it is easy to obtain the OTA-C filter shown in Figure 6.12, which is the counterpart network of the gyrator–capacitor filter shown in Figure 6.8b.

6.3.2 SC Filters

In order to implement a ladder filter in the SC technique, let us consider the node voltage equations of the gyrator–capacitor circuit from Figure 6.8b in the form

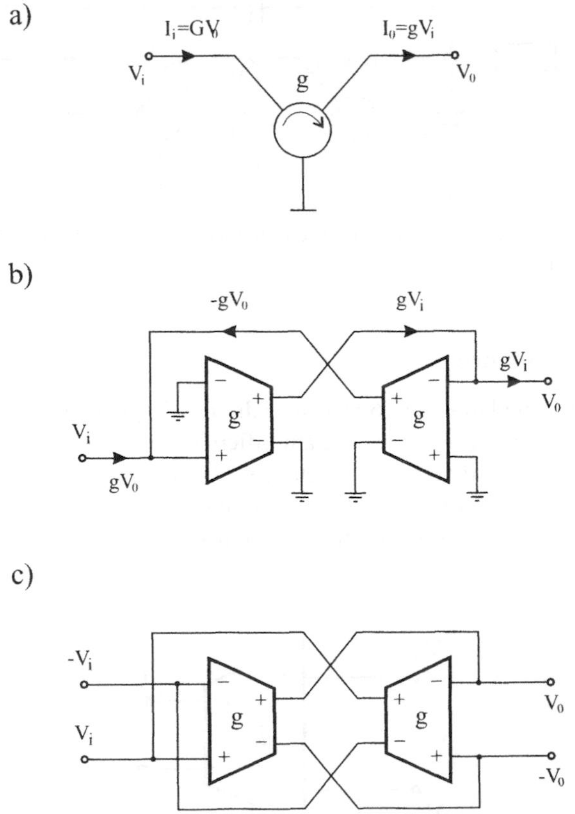

Figure 6.10 Implementations (b) and (c) of ideal gyrator (a) with the use of transconductance amplifiers.

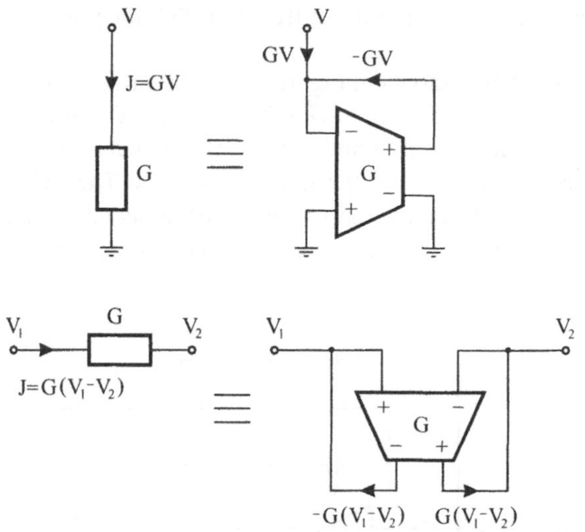

Figure 6.11 Implementation of conductances based on transconductance amplifiers.

$$\begin{bmatrix} s(C_1+C_2)+G_i & g & -sC_2 \\ -g & sC & g \\ -sC_2 & -g & s(C_2+C_3)+G_o \end{bmatrix} \cdot \begin{bmatrix} V_1 \\ V_2 \\ V_3 \end{bmatrix} = \begin{bmatrix} G_iV_i \\ 0 \\ 0 \end{bmatrix} \tag{6.28}$$

which can be written as follows

$$V_1 = \frac{-1}{s(C_1+C_2)+G_i}[-G_iV_i + gV_2 - sC_2V_3]$$

$$V_2 = \frac{-1}{sC}[-gV_1 + gV_3]$$

$$V_3 = \frac{-1}{s(C_2+C_3)+G_o}[-sC_2V_1 - gV_2] \tag{6.29}$$

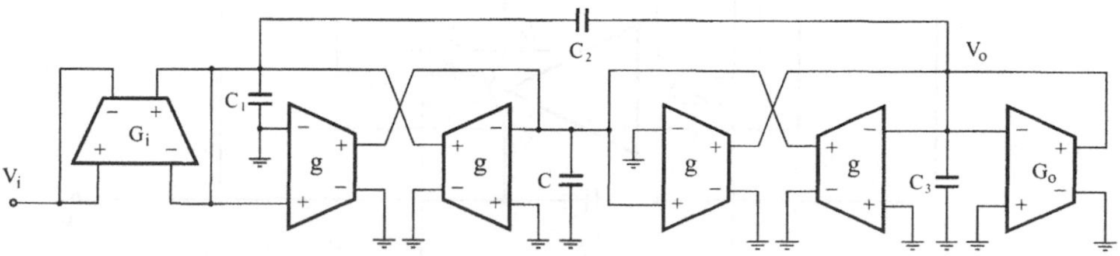

Figure 6.12 OTA-C implementation of a ladder prototype filter on the basis of the gyrator–capacitor circuit.

and modeled with the use of the fully differential SC integrators shown in Figure 6.13.

The obtained SC filter is shown in Figure 6.13. The values of capacitances in this filter are: $C_g = gT/2$, $C_i = G_iT/2$, $C_o = G_oT/2$, $C_a = C_1 + C_2$, and $C_b = C_2 + C_3$. SC counterpart networks of input (G_i) and output (G_o) branches are included into the feedback branches of the first and the third integrators. The SC circuit shown in Figure 5.14c is used as the counterpart network of the conductances G_i and G_o in Figure 6.8b. Let us note that the same circuit is used in order to implement the con-

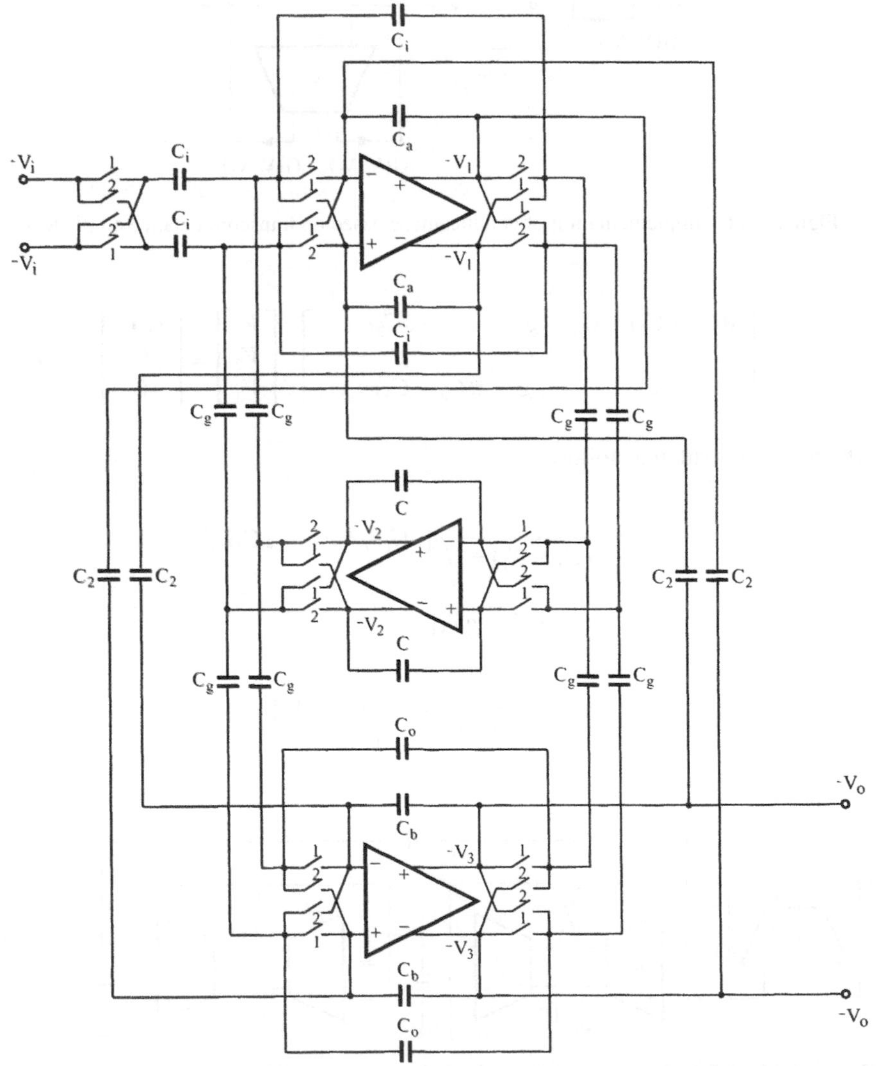

Figure 6.13 Third-order SC filter composed of fully differential amplifiers.

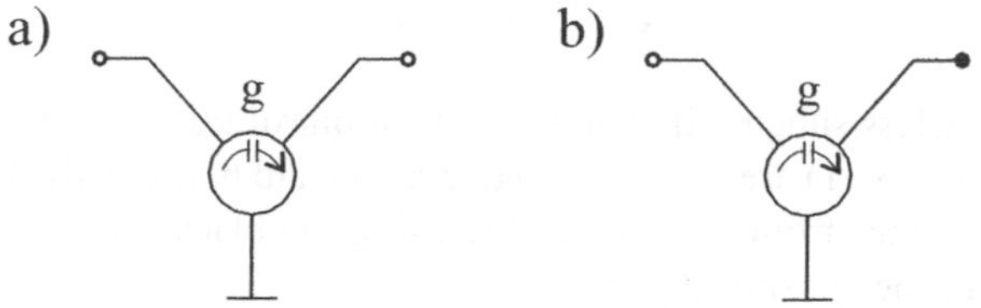

Figure 6.14 Symbols of quasiideal gyrators.

ductance G_i in the input branch of the first integrator and the gyrator transconductances g.

Equations (6.28) and (6.29) explain the design method of the SC counterpart network obtained on the basis of the gyrator–capacitor prototype network. However, comparison of circuits in Figure 6.8b and in Figure 6.13 shows that the structure of the SC counterpart network is determined by the prototype circuit structure and can be automatically generated from this circuit.

It was shown in Chapter 5 that SC circuits realized in a balanced structure are very advantageous for compensating for parasitic effects. However, in this structure a double number of capacitors is necessary, which increases the chip area. Hence, attention of designers is drawn to methods in which the Euler integrators shown in Figure 5.12 are used [36, 37]. One of these methods is based on prototype circuits containing quasiideal gyrators [19]. Quasiideal gyrators can be defined by the admittance matrices in the form

$$Y(s) = \begin{bmatrix} 0 & y(s) \\ -y(-s) & 0 \end{bmatrix} \tag{6.30}$$

or

$$Y(s) = \begin{bmatrix} 0 & y(-s) \\ -y(s) & 0 \end{bmatrix} \tag{6.31}$$

where

$$y(s) = g(1 + sT/2) \tag{6.32}$$

The matrices (6.30) and (6.31) can be written as a sum of two matrices

$$Y(s) = \begin{bmatrix} 0 & g \\ -g & 0 \end{bmatrix} \pm s\frac{gT}{2}\begin{bmatrix} 0 & 1 \\ 1 & 0 \end{bmatrix} \tag{6.33}$$

which means that the considered gyrator can be composed of the ideal gyrator g and the mutual capacitance $C_m = \pm gT/2$ in parallel connection. Hence, according to the definition of passivity (5.1), the subcircuit composed of a quasiideal gyrator terminated at its ports by capacitances C_1 and C_2 so that the condition

$$C_1C_2 - C_m^2 \geq 0 \tag{6.34}$$

is fulfilled is a lossless subcircuit. The symbols of quasiideal gyrators described by matrices (6.30) and (6.31) are shown in Figure 6.14a and b, respectively. We differentiate the first gyrator from the second one using the black node at the output of the second kind of the quasiideal gyrator.

Let us also note that the power P delivered to the ports of a quasiideal gyrator with a sinusoidal excitation is equal to zero. On the basis of (6.30) and (6.31) we have

$$I_1 = g(1 \pm j\omega T/2)V_2, \qquad I_2 = -g(1 \mp j\omega T/2)V_1 \tag{6.35}$$

Hence,

$$\begin{aligned} P = re[S(j\omega)] &= re[V_1(j\omega)I_1^*(j\omega) + V_2(j\omega)I_2^*(j\omega)] \\ &= re[V_1(j\omega)g(1 \mp j\omega T/2)V_2^*(j\omega) \\ &\quad -V_2(j\omega)g(1 \pm j\omega T/2)V_1^*(j\omega)] = 0 \end{aligned} \tag{6.36}$$

where the asterisk denotes a mutual number.

Using the bilinear transformation

$$\frac{sT}{2} = \frac{1 - z^{-1}}{1 + z^{-1}} \tag{6.37}$$

and introducing into the transfer function (5.17) the admittances $y(s)$ or $-y(-s)$ instead of G, we obtain

$$H(z) = -\frac{gT}{C}\frac{1}{1 - z^{-1}} \tag{6.38}$$

and

$$H(z) = \frac{gT}{C}\frac{z^{-1}}{1 - z^{-1}} \tag{6.39}$$

For $C_1 = gT$ the equations (6.38), (6.39) are equivalent to the equations (5.29), (5.33), respectively. Hence, the prototype circuit containing quasiideal gyrators can be bilinearly transformed into the discrete domain with the use of Euler integrators, as shown in Figure 5.12.

As an example, let us consider the circuit presented in Figure 6.8c, which for $C_2' = C_2 + C_L$, $C_L = T^2/(4L)$ and $C = g^2L$ is equivalent to the third-order reactance ladder filter from Figure 6.8a. The node voltage equations of the gyrator–capacitor circuit from Figure 6.8c have the form

$$\begin{bmatrix} s(C_1 + C_2') + G_i & y(s) & -sC_2' \\ -y(-s) & sC & y(-s) \\ -sC_2' & -y(s) & s(C_2' + C_3) + G_o \end{bmatrix} \cdot \begin{bmatrix} V_1 \\ V_2 \\ V_3 \end{bmatrix} = \begin{bmatrix} G_iV_i \\ 0 \\ 0 \end{bmatrix} \tag{6.40}$$

They can be written as follows

$$V_1 = \frac{-1}{s(C_1 + C_2') + G_i}[G_i(-V_i) + y(s)V_2 + sC_2'(-V_3)]$$

$$V_2 = \frac{-1}{sC}[-y(-s)V_1 - y(-s)(-V_3)]$$

$$-V_3 = \frac{-1}{s(C_2' + C_3) + G_o}[sC_2'V_1 + y(s)V_2] \tag{6.41}$$

and modeled with the use of the Euler SC integrators shown in Figure 5.12. The obtained SC filter is shown in Figure 6.15.

The SC counterpart network of the input branch G_i and the output branch G_o, shown in Figure 5.14b, is included into the feedback branch of the first integrator and into feedback branch of the third integrator. The SC circuit shown in Figure 5.14a is included into the input branch of the first integrator. The negative capacitance $-GT/2$ presented in Figure 5.14b is absorbed in the feedback capacitances of the integrators. Hence, the values of capacitances in the filter shown in Figure 6.15 are: $C_g = gT$, $C_i = G_iT$, $C_o = G_oT$, $C_a = C_1 + C_2' - C_i/2$ and $C_b = C_2' + C_3 - C_o/2$.

Similar to the method based on the prototype circuit composed of ideal gyrators, the structure of the SC filter shown in Figure 6.15 can also be automatically generated from the structure of a prototype circuit containing quasiideal gyrators. Let us note that the output voltage of the third integrator has the opposite sign with respect to the voltage V_3 in the prototype circuit. This sign change is marked in the proto-

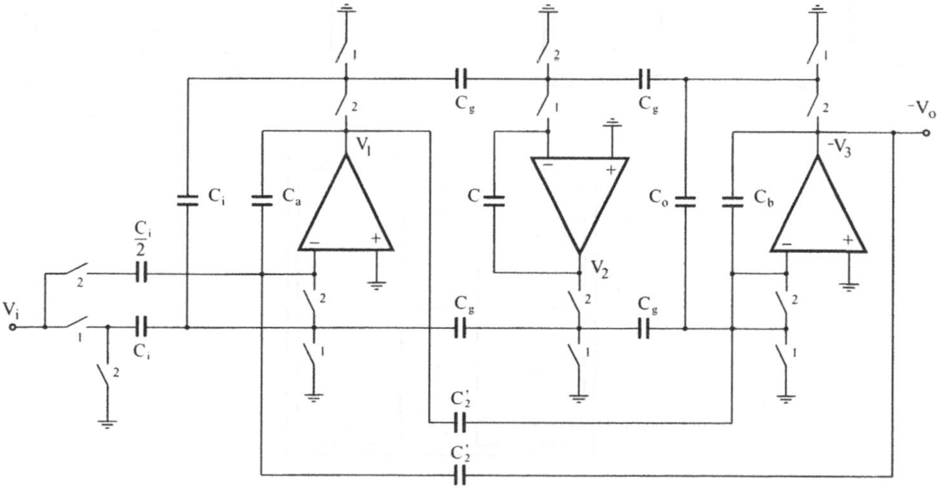

Figure 6.15 Third-order bilinear SC filter composed of Euler integrators.

type circuit presented in Figure 6.8c by the black node at the output port of the quasi-ideal gyrator described by (6.31).

There are several synthesis methods for circuits containing ideal gyrators [44]. Each circuit composed of ideal gyrators can be transformed into a circuit composed of quasiideal gyrators and vice versa with the use of relation (6.33). However, it is possible to modify the synthesis methods known for ideal gyrators in order to ob-

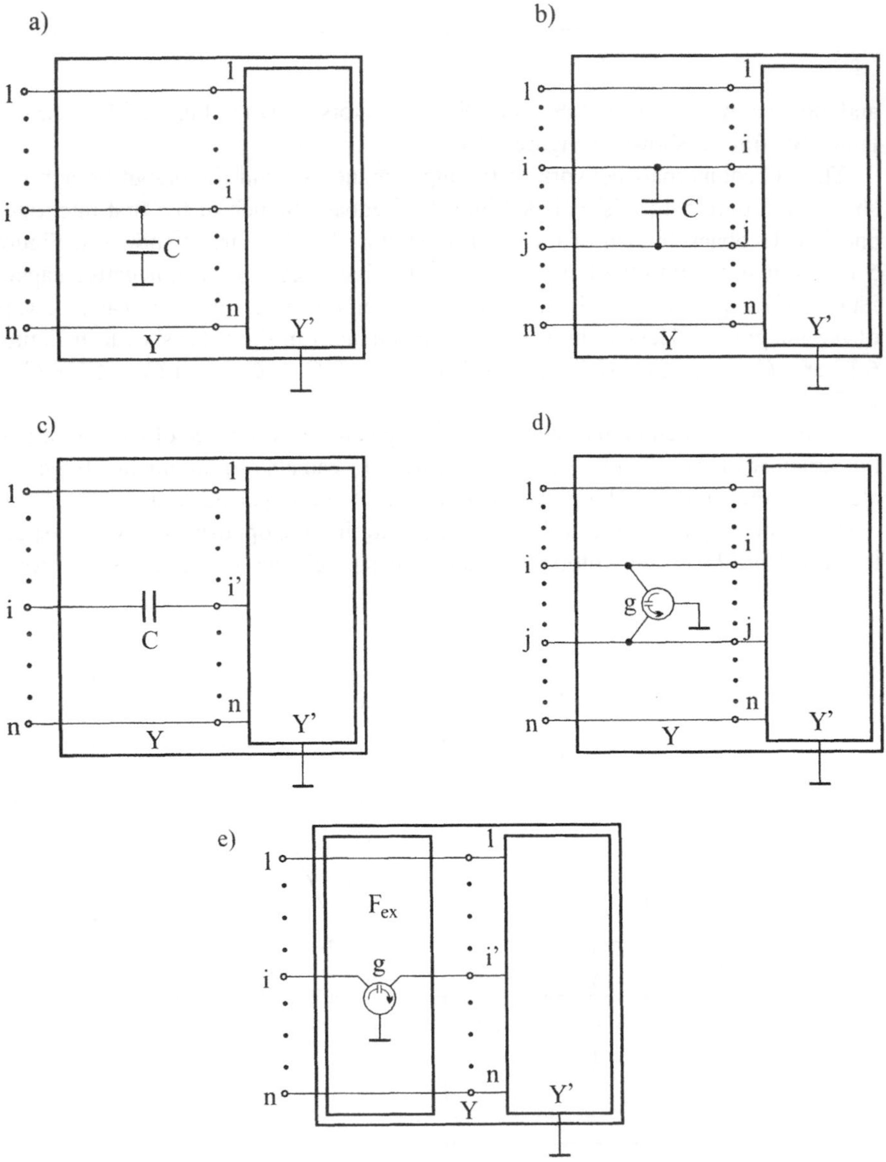

Figure 6.16 Extraction of capacitors: a) EC(i), b) PEC(i, j), c) SEC(i, i'); and gyrators: d) PEG(i, j), and e) SEG(i, i').

tain directly the multiport networks containing quasiideal gyrators. One of these methods, synthesis by extraction, is presented in [19].

Figure 6.16a,b,c shows the extraction of capacitors from a multiport network described by an admittance matrix. Figure 6.16d,e shows the parallel and series extraction of a gyrator. We will present formulae for the admittance matrix of residual multiport networks obtained after the chosen extraction procedure. Let us assume that the admittance matrix of a multiport network has the form

$$Y(s) = \frac{1}{m(s)}[l_{ij}(s)] \tag{6.42}$$

and that the polynomial $l^i_{jk}(s)$ fulfils the relation

$$ml^i_{jk} = l_{ii}l_{jk} - l_{ik}l_{ji} \tag{6.43}$$

In case of a two-port network, the polynomial l^i_{jk}, for $i, j, k = 1, 2$, is its characteristic polynomial n_{oo} [59].

Assuming that the admittance matrix $Y'(s)$ of a network obtained after extraction procedures is composed of polynomials m' and l'_{ij}, we describe the procedures by the following algebraic operations:

1. $EC(i)$—extraction of the grounded capacitor from the port i: $l'_{ii} = l_{ii} - msC$
2. $PEC(i, j)$—extraction of the parallel capacitor between the ports i, j: $l'_{ii} = l_{ii} - msC$, $l'_{ij} = l_{ij} + msC$, $l'_{ji} = l_{ji} + msC$, $l'_{jj} = l_{jj} - msC$
3. $SEC(i, i')$—extraction of the series capacitor from the port i: $m' = m - l_{ii}/sC$, $l'_{jk} = l_{jk} - l^i_{jk}/sC$, $j, k = 1, \ldots, n$, $j, k \neq i$
4. $PEG_{\pm}(i, j)$—extraction of the parallel gyrator between the ports i, j: $l'_{ij} = l_{ij} - g(1 \pm sT/2)m$, $l'_{ji} = l_{ji} + g(1 \mp sT/2)m$
5. $SEG_{\pm}(i, i')$—extraction of the series gyrator from the port i: $m' = l_{ii}$, $l'_{ii} = g^2(1 - s^2T^2/4)m$, $l'_{ij} = -g(1 \mp sT/2)l_{ij}$, $l'_{ji} = g(1 \pm sT/2)l_{ji}$, $l'_{jk} = l^i_{jk}$, $j, k = 1, \ldots, n$, $j, k \neq i$

The polynomials of the matrix Y', which do not change after the extraction procedures and are the same as in the matrix Y, are not listed in the above formulae. In the last two extraction procedures concerning the gyrator, the upper sign is chosen for the gyrator described by matrix (6.30), the lower one by (6.31).

It is easy to prove the formulae in the first, second, and fourth extraction procedures on the basis of the equation

$$Y = Y_{\text{ex}} + Y' \tag{6.44}$$

where Y_{ex} denotes the admittance matrix of the $2n$-port network containing the extracted element.

In order to prove the relations concerning the third and fifth extraction procedures, we can use the chain matrix F_{ex} of the extracted $2n$-port network in the form

$$F_{ex} = \begin{bmatrix} A_{ex} & B_{ex} \\ C_{ex} & D_{ex} \end{bmatrix} \tag{6.45}$$

where the dimensions of the matrices A_{ex}, B_{ex}, C_{ex}, and D_{ex} are $n \cdot n$ [44].

On the basis of the chain matrix of the extracted $2n$-port network, the relations between voltages and currents on n input and n output ports can be written in the form

$$V_1 = A_{ex}V_2 + B_{ex}I_2, \qquad I_1 = C_{ex}V_2 + D_{ex}I_2 \tag{6.46}$$

where V_1 and V_2 denote column matrices of the voltages at inputs and outputs of the $2n$-port network, respectively. Similarly, I_1 denotes the column matrix of the currents flowing into the inputs and I_2 denotes the column matrix of currents flowing from the outputs of the $2n$-port network. These voltages and currents are also related by the matrices Y and Y' in the following way:

$$I_1 = YV_1, \qquad I_2 = Y'V_2 \tag{6.47}$$

Hence, from (6.46) and (6.47) we obtain the equation

$$Y' = (D_{ex} - YB_{ex})^{-1}(YA_{ex} - C_{ex}) \tag{6.48}$$

It is easy to determine the matrices A_{ex}, B_{ex}, C_{ex}, and D_{ex} for the extraction shown in Figure 6.16c and for the one shown in Figure 6.16e. For example, for the series extraction of a gyrator shown in Figure 6.16e the matrices A_{ex} and D_{ex} have the unit elements on the diagonals, except the elements in the ith row and the ith column, which are equal to zero. The remaining elements of these matrices are also equal to zero. All elements of the matrices B_{ex} and C_{ex} are also equal to zero, except the elements in the ith row and the ith column. The nonzero element in the matrix B_{ex} is equal to $[g(1 \mp sT/2)]^{-1}$, whereas the nonzero element in the matrix C_{ex} is equal to $g(1 \pm sT/2)$. Introducing these matrices into (6.48) gives, after algebraic operations, the previously shown formulae for the series extraction procedure of a gyrator.

The method of synthesis by extraction consists in decreasing the degree of the admittance matrix complexity. The procedures of series extraction of a gyrator or a capacitor are helpful in realizing this goal. Let us consider, for example, a series extraction of a gyrator. Let us assume that the matrix Y has the factor $(1 \mp sT/2)$ in each numerator polynomial of the ith column and the factor $(1 \pm sT/2)$ in each numerator polynomial of the ith row. In order to fulfil this assumption, extraction of the grounded capacitor, extraction of the parallel capacitor, or extraction of the parallel gyrator can be used. If this assumption is fulfilled, then the series extraction of a gyrator, according to formulae given for $SEG_{\pm}(i, i')$, introduces the common factor $(1 - s^2T^2/4)$ in all polynomials of the matrix Y'. The common factor can be removed and the matrix complexity is decreased. This result can be formulated as a theorem.

Theorem. *If there exists an i such that for each* $j = 1, \ldots, n$ *the factor* $(1 \mp sT/2)$ *is the divisor of the polynomial* l_{ji} *and the factor* $(1 \pm sT/2)$ *is the divisor of the polynomial* l_{ij}, *then the procedure* $SEG_{\pm}(i, i')$ *decreases the complexity of the matrix Y.*

In order to illustrate the synthesis method based on extraction procedures, let us consider a low-pass filter with the transfer function

$$H(s) = \frac{s^2 - s + 1}{2(s^2 + s + 1)(s + 1)} \tag{6.49}$$

The numerator of the transfer function of a ladder reactance filter must be a purely even or odd polynomial [55]. Hence, the transfer function (6.49) cannot be obtained on the basis of the ladder reactance filter. However, this transfer function fulfils the relation $|H(j\omega)| \leq 1$ and, in accordance with (6.12), (6.14), and (6.15), can be realized by a lossless two-port network terminated by conductances $G_i = G_o = 1$. Hence, on the basis of (6.20) and (6.21), we have

$$S = \frac{1}{s^3 + 2s^2 + 2s + 1}\begin{bmatrix} -s^3 + s^2 - s & s^2 + s + 1 \\ s^2 - s + 1 & -s^3 - s^2 - s \end{bmatrix} \tag{6.50}$$

We see that this scattering matrix is nonsymmetrical, which is the consequence of the previously mentioned form of the numerator of the transfer function. In order to realize the transfer function with the use of a nonreciprocal lossless circuit, let us calculate the admittance matrix $Y = 2(S + 1)^{-1} - 1$. We have

$$Y = \frac{1}{s}\begin{bmatrix} s^2 + 1 & -s^2 - s - 1 \\ -s^2 + s - 1 & 3s^2 + 1 \end{bmatrix} \tag{6.51}$$

Using the procedure $PEC(1, 2)$: $C_1 = 7/4$, we obtain

$$Y = \frac{1}{s}\begin{bmatrix} -3s^2 + 1 & (1 - \frac{1}{2}s)(-\frac{3}{2}s - 1) \\ (1 + \frac{1}{2}s)(\frac{3}{2}s - 1) & \frac{5}{4}s^2 + 1 \end{bmatrix} \tag{6.52}$$

and next, using the procedure $EC(2)$: $C_2 = 3/2$, we obtain

$$Y = \frac{1}{s}\begin{bmatrix} -\frac{3}{4}s^2 + 1 & (1 - \frac{1}{2}s)(-\frac{3}{2}s - 1) \\ (1 + \frac{1}{2}s)(\frac{3}{2}s - 1) & 1 - \frac{1}{4}s^2 \end{bmatrix} \tag{6.53}$$

which fulfils the given theorem. The complexity degree of Y can be decreased with the use of the procedure $SEG_+(2, 2')$: $g = 1$. Hence, the residual two-port network has the following admittance matrix:

$$Y = \begin{bmatrix} \frac{3}{2}s & -\frac{3}{2}s - 1 \\ -\frac{3}{2}s + 1 & s \end{bmatrix} \tag{6.54}$$

Using again the extraction procedures $PEC(1, 2')$: $C_3 = 1$, $PEG_-(2', 1)$: $g = 1$, and $EC(1)$: $C_4 = 1/2$ we get the residual two-port network with the matrix $Y = 0$.

The prototype circuit realizing the transfer function (6.49) is shown in Figure 6.17a, where the gyrators g_1 and g_2 are described by the matrices (6.30) and (6.31), respectively. The circuit has the node voltage equations in the form

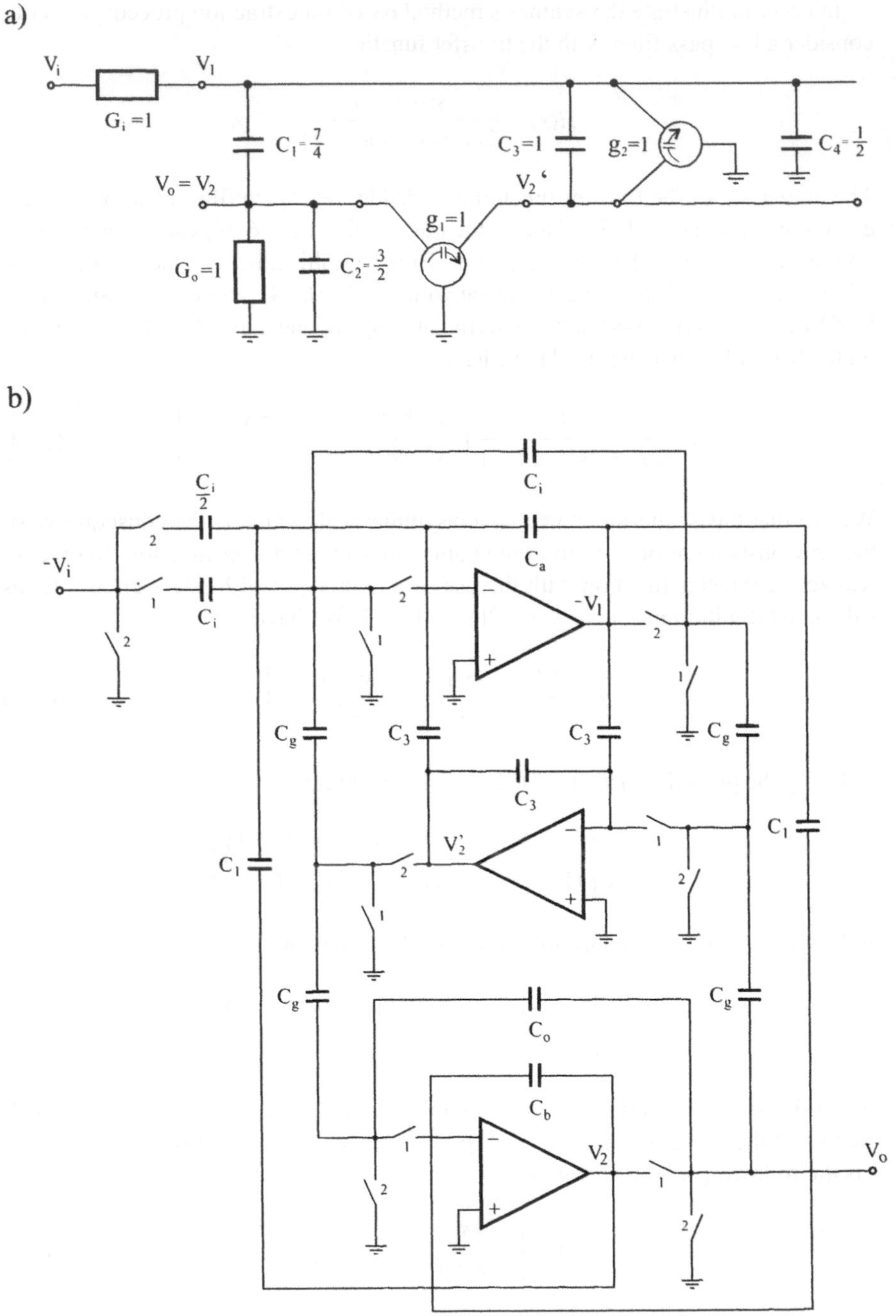

Figure 6.17 Prototype low-pass filter composed of quasiideal gyrators (a), and its SC counterpart network (b).

$$\begin{bmatrix} s(C_1 + C_3 + C_4) + G_i & -sC_1 & -y(s) - sC_3 \\ -sC_1 & s(C_1 + C_2) + G_o & -y(-s) \\ y(-s) - sC_3 & y(s) & sC_3 \end{bmatrix} \cdot \begin{bmatrix} V_1 \\ V_2 \\ V_2' \end{bmatrix} = \begin{bmatrix} G_i V_i \\ 0 \\ 0 \end{bmatrix} \tag{6.55}$$

which can be written as follows

$$-V_1 = \frac{-1}{s(C_1 + C_3 + C_4) + G_i}[G_i V_i + sC_1 V_2 + y(s)V_2' + sC_3 V_2']$$

$$V_2 = \frac{-1}{s(C_1 + C_2) + G_o}[sC_1(-V_1) - y(-s)V_2']$$

$$V_2' = \frac{-1}{sC_3}[sC_3(-V_1) - y(-s)(-V_1) + y(s)V_2] \tag{6.56}$$

On the basis of equations (6.56), we obtain the SC counterpart network shown in Figure 6.17b. In order to ensure the coincidence of the clocks that control the switches, the admittance $y(s)$ in the first and third equations of (6.56) has to correspond with the integrators described in rows 11 and 12 of Table 5.1, and the admittance $-y(-s)$ in the second and third equations has to correspond with integrators 6 and 3, respectively. For the unit value of the clock period $T = 1$, the capacitance values of the SC filter are as follows: $C_i = G_i T = 1$, $C_o = G_o T = 1$, $C_g = gT = 1$, $C_1 = 7/4$, $C_3 = 1$, $C_a = C_1 + C_3 + C_4 - G_i T/2 = 11/4$, $C_b = C_1 + C_2 - G_o T/2 = 11/4$. Hence, we have the capacitance spread $\Sigma C_n/C_{\min} = 39$.

For comparison, let us realize the transfer function (6.49) with the use of the prototype circuit in Figure 6.18a, composed of ideal gyrators. The circuit has the node voltage equations in the form

$$\begin{bmatrix} s+1 & -s-1 & 1 \\ -s+1 & 3s+1 & -1 \\ -1 & 1 & s \end{bmatrix} \cdot \begin{bmatrix} V_1 \\ V_2 \\ V_3 \end{bmatrix} = \begin{bmatrix} 1 \cdot V_i \\ 0 \\ 0 \end{bmatrix} \tag{6.57}$$

which can be written as follows

$$V_1 = \frac{-1}{s+1}[-1 \cdot V_i - (s+1)V_2 + 1 \cdot V_3]$$

$$V_2 = \frac{-1}{3s+1}[(-s+1)V_1 - 1 \cdot V_3]$$

$$V_3 = \frac{-1}{s}[-1 \cdot V_1 + 1 \cdot V_2] \tag{6.58}$$

The modeling of the above equations with the use of a fully differential integrator from Figure 6.13 gives the SC filter shown in Figure 6.18b. The capacitances in the filter are as follows: $C_1 = 1/2$, $C_2 = 1$, $C_3 = 1/2$, $C_4 = 1/2$, $C_5 = 1/2$, $C_6 = 1$, $C_7 = 1/2$, $C_8 = 1/2$, $C_9 = 1/2$, $C_{10} = 1/2$, $C_{11} = 1$, $C_{12} = 1$, $C_{13} = 3$, $C_{14} = 1/2$. Hence, we have the capacitance spread $\Sigma C_n/C_{\min} = 46$.

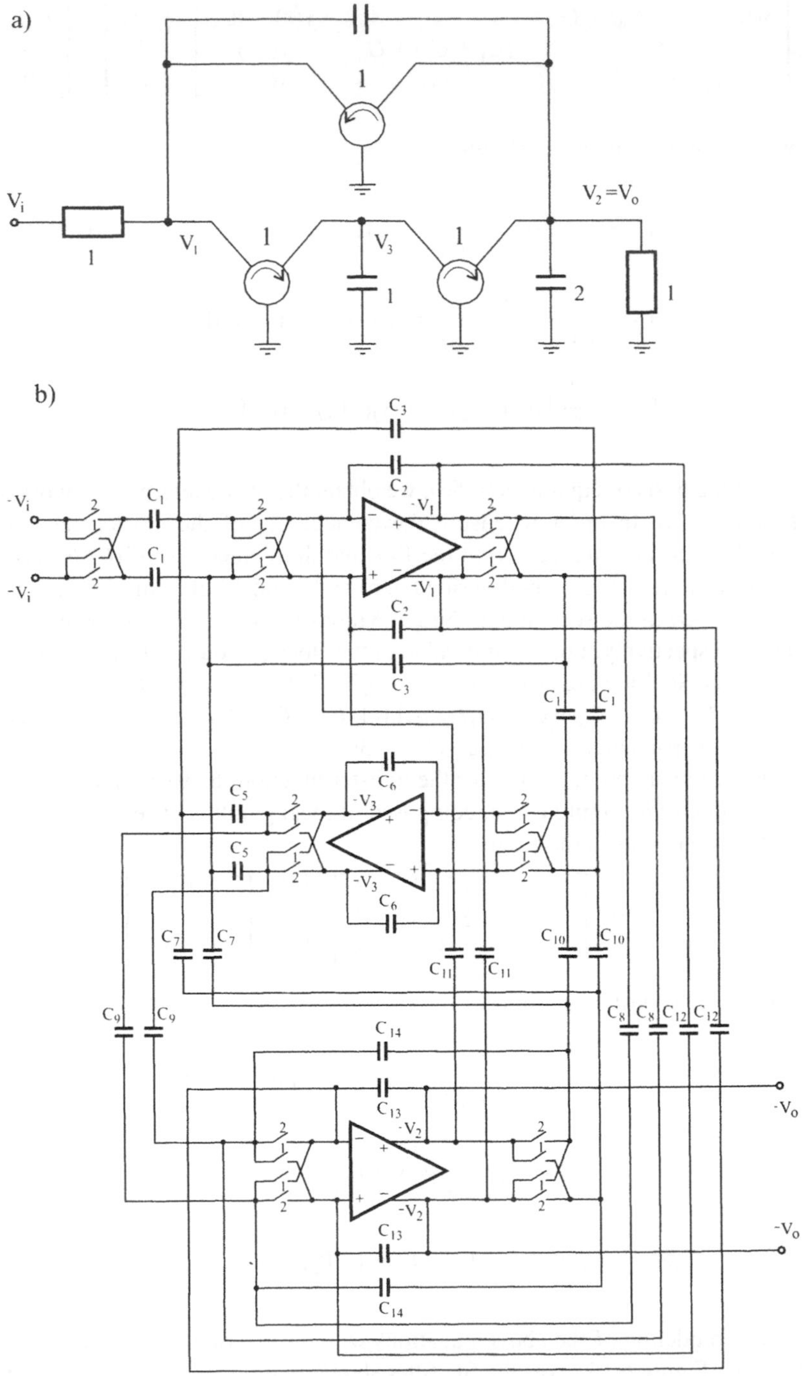

Figure 6.18 Low-pass filter prototype circuit composed of ideal gyrators (a), and its SC counterpart network (b).

Table 6.3 Sensitivity properties of a third-order balanced SC filter

	$0 \le \omega \le \pi/6$	
$\lvert S_{C_i}^{\lvert H\rvert}\rvert \le 0.3$	$\lvert S_{C_i}^{\lvert H\rvert}\rvert \le 0.7$	$\lvert S_{C_i}^{\lvert H\rvert}\rvert \le 1$
C_2, C_6, C_{11}, C_{12}	$C_3, C_7, C_8, C_{13}, C_{14}$	$C_1, C_4, C_5, C_9, C_{10}$
	$0 \le \omega \le \pi/2$	
$\lvert S_{C_i}^{\lvert H\rvert}\rvert \le 0.5$	$\lvert S_{C_i}^{\lvert H\rvert}\rvert \le 1$	$\lvert S_{C_i}^{\lvert H\rvert}\rvert \le 1.5$
$C_6, C_7, C_8, C_{11}, C_{14}$	$C_1, C_2, C_3, C_4, C_5, C_9, C_{10}$	C_{12}, C_{13}

The sensitivity of both filters has been analyzed with the use of topological methods. The analysis confirms the low sensitivity properties of active filters designed on the basis of lossless prototype circuits. Table 6.3 shows the results for the fully differential implementation. Assuming that $T = 1$ and using the prewarping relation (1.31), we have the limit frequency of the passband $\omega_0 = 0.464$ and the Nyquist frequency $\omega_N = \pi$. The capacitors are divided into classes, depending on the value of sensitivity with respect to the capacitance. The top part of the table shows sensitivity properties of the filter in the range of low frequencies ($0 \le \omega \le \pi/6$), and the bottom part refers to high frequencies ($0 \le \omega \le \pi/2$). The sensitivity of the transfer function magnitude $S_{C_i}^{|H|}$ with respect to capacitors C_2 and C_3 is shown in Figure 6.19.

6.3.3 Digital Filters

Digital filters can be implemented with the use of digital elements like adders/subtractors, multipliers, and registers composed of flip-flops, which were described in Chapter 4. With the use of these elements, the digital filter can be easily realized on the basis of its signal flow graph (SGF). In order to obtain the signal flow graph, we

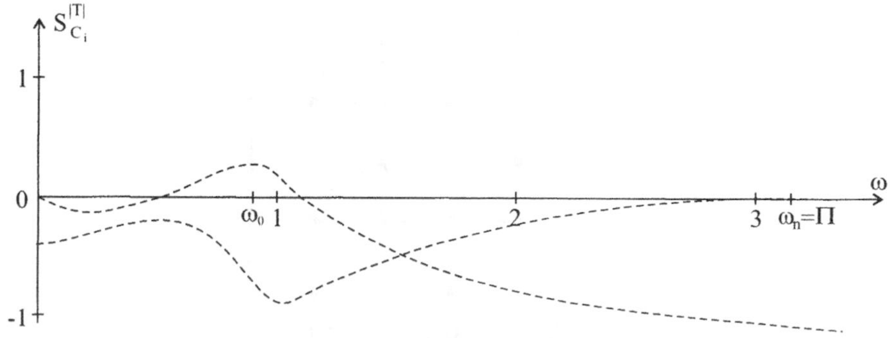

Figure 6.19 Sensitivity of the transfer function magnitude with respect to capacitors C_2 and C_3 for a fully differential SC filter.

can apply implicit or explicit approaches. We can obtain an SGF implicitly from a lossless prototype circuit or explicitly from the transfer function of a filter. In this section, we will describe both approaches and consider the sensitivity properties of the obtained filters.

First, let us describe the explicit approach. In digital filter design, the transfer function $H(z)$ is often obtained on the basis of the bilineary transformed transfer function in the analog domain s, and can written in the form

$$H(z) = h_0 \frac{1 + \sum_{i=1}^{M} \beta_i z^{-i}}{1 + \sum_{j=1}^{N} \alpha_j z^{-j}} = \frac{Y(z)}{X(z)} \tag{6.59}$$

where $X(z)$ is the excitation signal and $Y(z)$ is the response signal. Relation (6.59) implies the equations

$$Y(z) = \left(1 + \sum_{i=1}^{M} \beta_i z^{-i}\right) h_0 X(z) - \sum_{j=1}^{N} \alpha_j z^{-j} Y(z) \tag{6.60}$$

and

$$Y(z) = h_0 X(z) + z^{-1}\{[\beta_1 h_0 X(z) - \alpha_1 Y(z)] + z^{-1}\,[(\beta_2 h_0 X(z) - \alpha_2 Y(z)) + \cdots]\} \tag{6.61}$$

From equation (6.61) we can obtain the signal flow graph shown in Figure 6.20, called the direct form I structure [29]. The direct form II structure is obtained on the basis of the equations

$$\frac{Y(z)}{W(z)} = 1 + \sum_{i=1}^{M} \beta_i z^{-i} \tag{6.62}$$

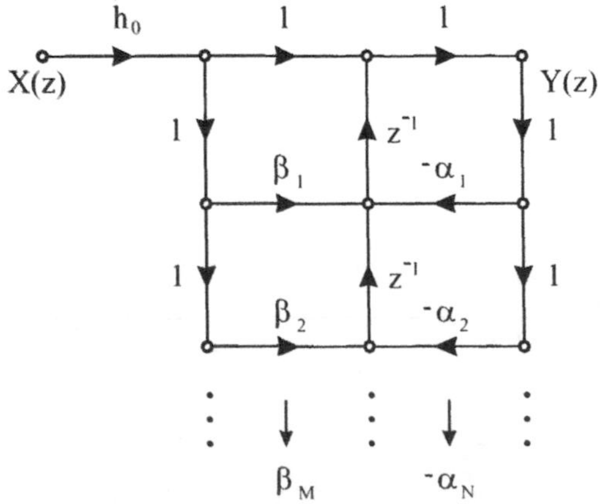

Figure 6.20 Direct form I structure of a digital filter.

and

$$\frac{X(z)}{W(z)} = \frac{1}{h_0}\left(1 + \sum_{j=1}^{N} \alpha_j z^{-j}\right) \tag{6.63}$$

which are equivalent to (6.59), and can be written in the form

$$Y(z) = W(z) + \sum_{i=1}^{M} \beta_i z^{-i} W(z) \tag{6.64}$$

$$W(z) = h_0 X(z) - \sum_{j=1}^{N} \alpha_j z^{-j} W(z) \tag{6.65}$$

giving the signal flow graph depicted in Figure 6.21.

Other structures can be obtained for the transfer function (6.59) decomposed into partial fractions

$$H(z) = H_1(z) + H_2(z) + \cdots H_L(z) \tag{6.66}$$

where $H_i(z)$, $i = 1, \ldots L$, are second-order terms. If the order N of the denominator polynomial in (6.59) is odd, one of the components is a first-order term. These terms can be realized in the form shown in Figure 6.20 or in Figure 6.21, giving the filter in the parallel form, depicted in Figure 6.22. Similarly, the cascade form of the filter structure, shown in Figure 6.23, corresponds to the transfer function decomposition

$$H(z) = H_1'(z) H_2'(z) \cdots H_L'(z) \tag{6.67}$$

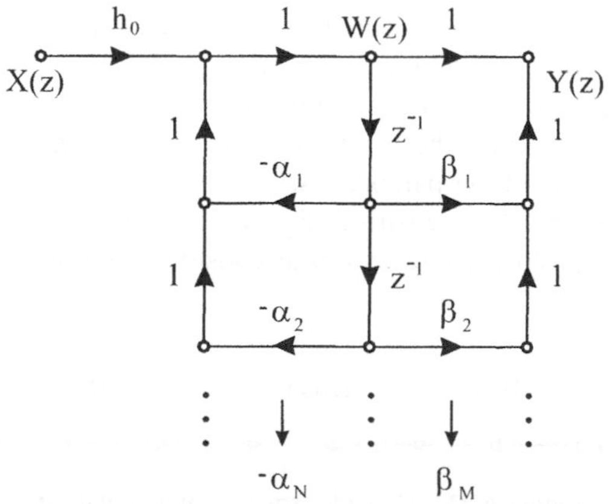

Figure 6.21 Direct form II structure of a digital filter.

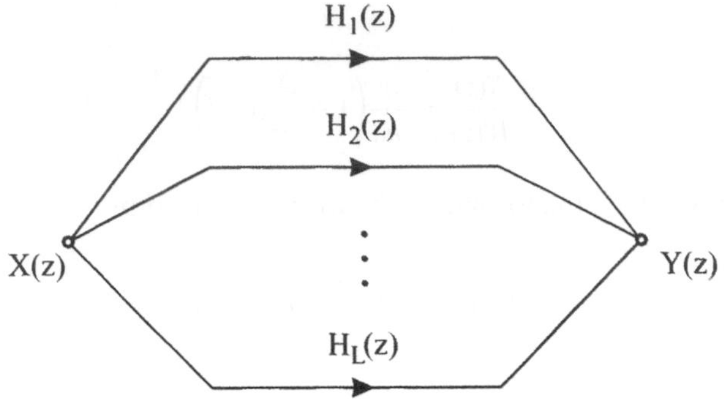

Figure 6.22 Parallel structure of a digital filter.

In order to illustrate the explicit approach, let us consider the filter with the transfer function (6.49). This transfer function will be transformed into the discrete domain with the use of relations $P_d = PT_3$ and $Q_d = QT_3$, taken from (1.57). The transformation matrix PT_3 is given in (1.49). Since the elements of transformation matrices were obtained on the assumption that the sampling period $T = 2$, the frequency in (6.49) should be also scaled with the use of the substitution $s \rightarrow 2s$, which gives

$$H(s) = \frac{4s^2 - 2s + 1}{2(8s^3 + 8s^2 + 4s + 1)} \tag{6.68}$$

Hence, for $P = [0\ \ 4\ \ -2\ \ 1]$, $Q = [8\ \ 8\ \ 4\ \ 1]$ we have $P_d = [7\ \ 1\ \ -3\ \ 3]$, $Q_d = [-3\ \ 15\ \ -25\ \ 21]$, respectively, and

$$H(z) = \frac{7z^{-3} + z^{-2} - 3z^{-1} + 3}{2(-3z^{-3} + 15z^{-2} - 25z^{-1} + 21)} = \frac{1 - z^{-1} + 1/3z^{-2} + 7/3z^{-3}}{14(1 - 25/21z^{-1} + 5/7z^{-2} - 1/7z^{-3})} \tag{6.69}$$

The direct form II structure of the obtained filter is shown in Figure 6.24.

In digital filters obtained explicitly from a transfer function, the parameters of the element are equal to coefficients of numerator and denominator polynomials of the transfer function. Hence, the magnitude and phase of the transfer function are sensitive to variations of filter parameters.

In order to realize a low-sensitive digital filter, we use an implicit approach based on a lossless prototype circuit. Let us consider the prototype circuit shown in

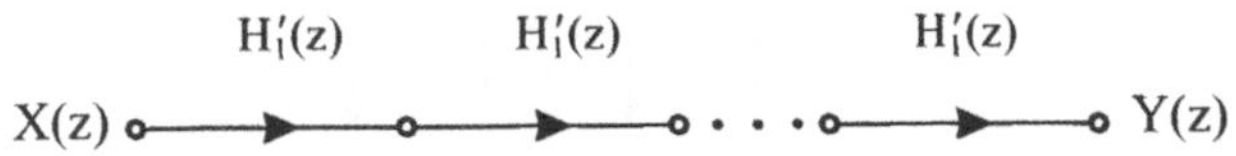

Figure 6.23 Cascade structure of a digital filter.

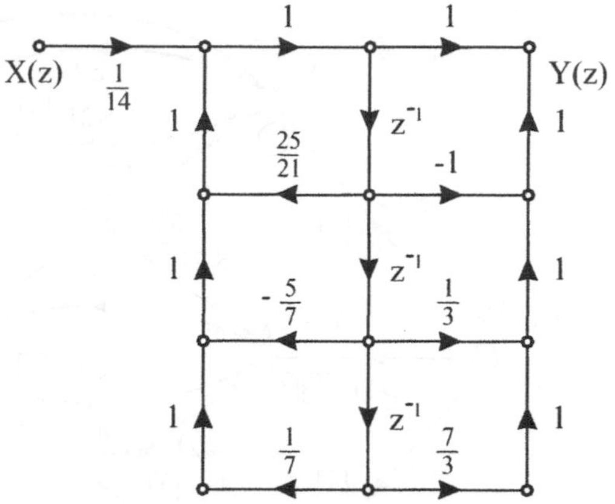

Figure 6.24 Example of a digital filter with the direct form II structure.

Figure 6.18a and described by the node voltage equations (6.57). For $T = 1$, after the bilinear transformation (6.37), these equations have the form

$$(A + z^{-1}B) \cdot \begin{bmatrix} V_1 \\ V_2 \\ V_3 \end{bmatrix} = \begin{bmatrix} (1 + z^{-1})\, V_i \\ 0 \\ 0 \end{bmatrix} \tag{6.70}$$

where

$$A = \begin{bmatrix} 3 & -3 & 1 \\ -1 & 7 & -1 \\ -1 & 1 & 2 \end{bmatrix}, B = \begin{bmatrix} -1 & 1 & 1 \\ 3 & -5 & -1 \\ -1 & 1 & -2 \end{bmatrix} \tag{6.71}$$

In order to eliminate loops without delays, the above equations can be written in the form

$$\begin{bmatrix} V_1 \\ V_2 \\ V_3 \end{bmatrix} = -z^{-1}A^{-1}B \begin{bmatrix} V_1 \\ V_2 \\ V_3 \end{bmatrix} + (1 + z^{-1})\, V_i \begin{bmatrix} 5/14 \\ 1/14 \\ 1/7 \end{bmatrix} \tag{6.72}$$

where

$$A^{-1}B = \begin{bmatrix} 5/21 & -4/7 & 8/21 \\ 8/21 & -5/7 & -4/21 \\ -4/7 & 4/7 & -5/7 \end{bmatrix} \tag{6.73}$$

for which we obtain the signal flow graph shown in Figure 6.25. The problem of loops without delays is discussed in more detail in [5, 14]. Both signal flow graphs

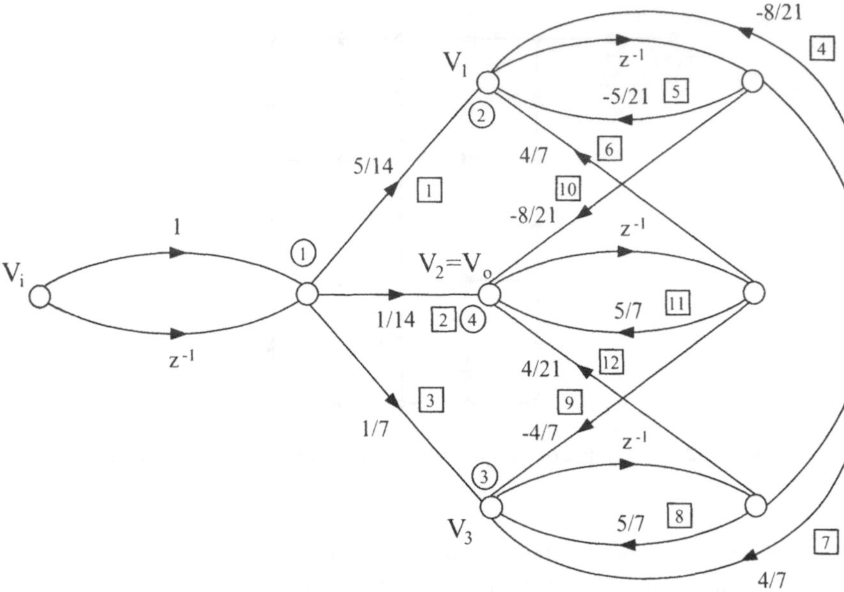

Figure 6.25 Signal flow graph of a digital filter obtained from a lossless prototype circuit.

differ in the number of multiplication (O_m) and addition/subtraction ($O_{a/s}$) operations that are necessary to realize the filter. For the graph in Figure 6.24, we have $O_m = 6$ and $O_{a/s} = 6$, whereas for the graph in Figure 6.25 $O_m = 12$ and $O_{a/s} = 10$. The number of additions/subtractions corresponding to each node of the graph is given by $O_{a/s} = b - 1$, where b is number of branches directed to the node.

A signal flow graph shows the algorithm of operation of a filter. However, it does not show the structure of the filter. As we mentioned earlier, a digital filter is assembled from adders, multipliers, and registers. However, we can expect that the number of elements necessary to implement the filter depends not only on the number of clock periods in which the filter operates but also on the signal flow graph used for this implementation.

In order to illustrate the hardware implementation of a digital filter on a chip, we consider the signal flow graph shown in Figure 6.25 and its equivalent form obtained after transformation of the graph, shown in Figure 6.26. Multiplications in the signal graph are numbered in boxes and the additions in circles. We assume that it is necessary to execute all operations in five clock cycles and that both the multiplier and the adder operate in one clock cycle. The input variable as well as state variables stored in the registers and available at their outputs can be used in the considered clock cycle. Operations that can be performed under these assumptions in each clock cycle, for both signal graphs, are shown in Table 6.4. The columns C, M, and A correspond to a clock cycle, multiplication and addition, respectively. We see from this table that the implementation of the signal flow graph "a" needs four mul-

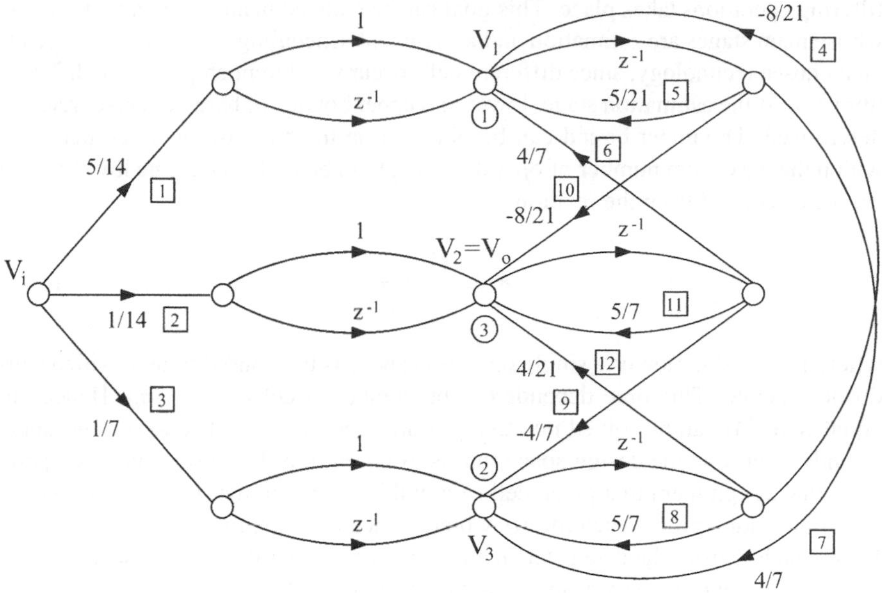

Figure 6.26 Transformed signal flow graph.

tipliers whereas the graph "b" requires only three multipliers. The number of adders necessary for implementation of graph "a" is also greater than for graph "b." This example shows that the signal flow graph transformation at the stage of behavioral synthesis is very important with respect to chip area minimization of the implemented filter.

In general, the task of hardware implementation of a filter is reduced to resource allocation with respect of the graph transformation obtained, and can be formulated as follows: for a given signal flow graph, maximum time t_{max} of filter operation and a given cell library, find the filter implementation with the minimal area on a chip. Usually, t_{max} is determined by the number of clock periods in which one cycle of

Table 6.4 Execution of operations for different signal flow graphs of a digital filter

	SGFa		SGFb	
C	M	A	M	A
1		1	1, 2, 3	
2	1, 2, 3, 4		4, 5, 6	
3	5, 6, 7, 8		7, 8, 9	1
4	9, 10, 11, 12	2	10, 11, 12	2
5		3, 4		3

filtering operations takes place. This goal can be realized in an optimization process whose main stages are estimation and assignment/scheduling. The solution depends on a chosen technology, since different cells occupy different chip areas in different libraries. At the estimation stage, lower and upper bounds of hardware resources are determined. The upper bound can be obtained on the basis of the clock period in which the maximum number of operations ought to be performed. The lower bound can be calculated from the relation

$$n_i \geq \frac{O_i t_i + t_u}{t_{\max}} \tag{6.74}$$

where t_i is the duration of a single operation and t_u is the unused time in which cells do not operate. This time depends on the number of cells, $t_u = t_u(n_i)$. Hence, inequality (6.74) can be solved iteratively, starting with $t_u = 0$. The lower and upper bounds determine the design space that is explored in order to schedule the operations and assign them to a given cell of a multiplier or summer [4]. In other words, the optimization process determines which particular cell realizes a given operation, from which register the data is taken and to which register the result is sent. In order to control these processes, a finite-state machine is used.

6.3.4 SI Filters

It was shown in the first part of this book that switched-current circuits operate in the current mode. In other words, signals correspond to currents, exclusively. According to the Kirchhoff's current law, the adder has a very simple implementation as a node in a circuit. The current mirror and the memory cell shown in Figure 5.2a and Figure 5.5, respectively, can be used as multipliers and delay elements in a signal flow graph. Hence, a signal flow graph can be directly implemented with the use of such elements. A bilinear integrator depicted in Figure 5.8 can serve as an example of an SI circuit obtained on the basis of a signal flow graph.

SI filters can also be implemented using methods introduced for switched-capacitor circuits. Let us consider the method presented in [1], originally introduced for switched-capacitor filters [30]. As an example, we take the gyrator–capacitor circuit shown in Figure 6.8b. The node voltage equations (6.28) of the prototype circuit can be written as follows

$$V_1 = \frac{1}{s(C_1 + C_2)}[G_i V_i - G_i V_1 - gV_2] + \frac{C_2}{C_1 + C_2} V_3$$

$$V_2 = \frac{g}{sC}[V_1 - V_3]$$

$$V_3 = \frac{1}{s(C_2 + C_3)}[-G_o V_3 + gV_2] + \frac{C_2}{C_2 + C_3} V_1 \tag{6.75}$$

and modeled with the use of the fully differential SI integrators shown in Figure 5.7 or in Figure 5.8. The obtained SI filter is shown in Figure 6.27, where two-input blocks denote integrators and one-input blocks denote current mirrors. Using the integrator presented in Figure 5.8, we calculate the parameters α_i and β_i for $i = 0,1$ from formulae (5.26) and (5.27). In these calculations, we have the damping factor of all integrators $\alpha = 0$. The coefficient a_1 is $a_1 = 1/(C_1 + C_2)$ for the first integrator, $a_1 = g/C$ for the second integrator, and $a_1 = 1/(C_2 + C_3)$ for the third one. Voltage signals from the prototype circuit correspond to currents in its SI counterpart net-

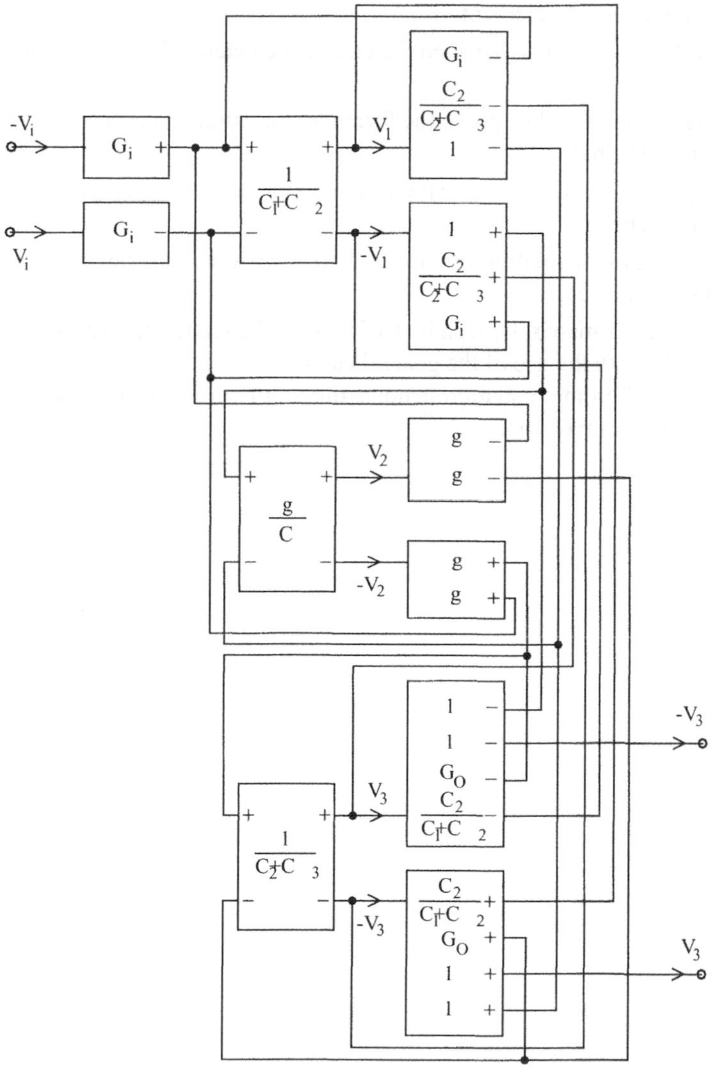

Figure 6.27 Switched-current implementation of a lossless prototype circuit.

work. In this approach, feedback capacitors in the prototype circuit are replaced by current feedbacks with the coefficients equal to capacitor ratios.

6.4 PROBLEMS

1. Calculate the signals v_i, y_i, and f_i in the delta-sigma modulator shown in Figure 6.2, for $x_{\min} = -4$, $x_{\max} = 4$, and $x_k = -1$. Compare the results with the values shown in Table 6.1.
2. With the use of definition (5.1) show that the mutual capcitance described by (6.34) is a lossless element.
3. Show the equivalence between the circuits depicted in Figure 6.8a and Figure 6.8c.
4. Using equation (6.44), prove the formulae for extraction procedures shown in Figure 6.16a,b,d.
5. Using equation (6.48), prove the formulae for extraction procedures shown in Figure 6.16c,e.
6. On the basis of relations (6.46) and (6.47), show that the matrix equation (6.48) is fulfilled.
7. Calculate the matrix Y given by (6.51) from the scattering matrix S described by (6.50) with the use of the general equation $Y = 2(S + 1)^{-1} - 1$. Note that the symbol "1" in this equation denotes the unit matrix and the power "–1" denotes matrix inversion.

7

Image Processing

In this chapter we describe the architecture of a system on a chip that can be used for image processing. We assume that image sensing arrays are built on the chip in the standard CMOS process. Switched-current circuits are used for image preprocessing in the analog part of the chip.

7.1 INTRODUCTION

Together with standards and networking, VLSI circuits play an important role in multimedia applications. In this chapter, we will consider a one-chip implementation of algorithmic techniques (standards) that can be used in portable consumer products such as digital video cameras. The following techniques are used for video compression in the standards MPEG and H.26x:

1. Chrominance subsampling
2. Frame differencing
3. Motion compensation
4. Interpolation
5. Transform coding
6. Entropy coding

Compression is at the heart of every standard system. Uncompressed video sequences require more than 100 Mb/s and communication networks such as ISDN cannot transmit so much data. However, both human perception and information theory imply redundancy in video data. The techniques enumerated above, which exploit this redundancy, lead to significant compression.

Color images and video can be represented in different color spaces. For images generated in image sensing arrays or displayed on screens, red (R), green (G), and blue (B) components are used. However, there is a strong correlation between these components in natural images. In image processing, it is advisable to use a color space with luminance (Y) and chrominance (C_R) and (C_B) components defined as

$$\begin{bmatrix} Y \\ C_R \\ C_B \end{bmatrix} = \frac{1}{2} \begin{bmatrix} 0.598 & 1.174 & 0.228 \\ 1.000 & -0.838 & 0.162 \\ -0.338 & -0.662 & 1.000 \end{bmatrix} \cdot \begin{bmatrix} R \\ G \\ B \end{bmatrix} \tag{7.1}$$

The above relations reduce component correlation. With the use of logarithmic and nonlinear relations, other color spaces can be defined in which component correlation is further decreased. The advantage of luminance and chrominance space is spatial resolution reduction of the chrominance. Due to the properties of human perception, *chrominance subsampling* causes little loss in image quality. This subsampling is known as 4:1:1 coding, in which each 2×2 block of picture elements (pixels) is described by four luminance values, one value of C_R chrominance, and one value of C_B chrominance.

There is little variation in consecutive frames of a video sequence. Hence, instead of coding each frame separately, the difference between neighboring frames, *frame differencing*, is calculated for further processing.

Coding efficiency can also be improved significantly due to *motion compensation* [11]. Motion is caused either by moving objects in the frame or by a pan of the scene. In the previous frame, we search for a macroblock taken from the current frame. In the standards MPEG and H.26x, a macroblock consists of 16×16 luminance pixels and the search range is equal to ± 15 pixels in horizontal and vertical directions. For the best match, we code the difference of the macroblocks and the vector describing the offset of the macroblocks.

Further improvement of video coding efficiency is obtained using *interpolation*. In this technique, an intermediate frame is obtained as the weighted average of the previous and subsequent frames.

Transform coding is a very effective technique for image compression. The Karhunen–Loeve transform (KLT) is optimal for image compression because it packs the greatest amount of energy in the smallest number of elements in the frequency domain of a two-dimensional (2-D) signal and minimizes the total entropy of this signal sequence. However, the basis functions of KLT are image-dependent, which is the most important implementation-related deficiency. From other image-independent basis functions, the discrete cosine transform (DCT) basis is seen to be close to the output produced by the KLT. Hence, DCT-based image coding is the basis for all video compression standards. In these standards, the image is divided into 8×8 blocks in the spatial domain and DCT transforms them into 8×8 blocks in the 2-D frequency domain. Such block size is convenient with respect to computational complexity and larger size does not offer significantly better compression.

DCT implementation presented in this chapter will be based on relations (1.35) and (1.36).

According to information theory, small values of DCT data, which occur more frequently than large values, can be represented with the use of fewer bits. Huffman coding is an implementation of such *entropy coding* in the standards MPEG and H.26x.

The next section shows which of the above techniques are helpful in one-chip, portable video devices.

7.2 CMOS ONE-CHIP CIRCUIT FOR DIGITAL VIDEO CAMERAS

Image sensing arrays built with the standard CMOS process and composed of photodiodes or phototransistors are recent alternatives to the charge-coupled device (CCD) technology [40]. Hence, it is possible to reduce device complexity and to assemble the whole video system on a single CMOS chip. Implementations of selected techniques of the standards MPEG and H.26x in low-cost, low-power, one-chip portable devices like video cameras, are presented in recent papers [33, 52, 38]. In these implementations, the analog part of a mixed-signal chip is realized using the switched-capacitor technique. In this section, we will also present a switched-current implementation of transform and entropy coding on a chip containing an image sensing array.

The architecture of a one-chip digital video camera is presented in Figure 7.1. It contains an image sensor array divided into blocks consisting of 8 × 8 pixels each, a preprocessor section, a two-dimensional (2-D) discrete cosine transform (DCT) processor, an analog-to-digital converter, and an entropy coding section [33]. In order to obtain color images, the cells of the array are alternately covered by red, green, and blue (RGB) filters following the Bayer pattern, as shown in Figure 7.2 for an 8 × 8 pixel block. Preprocessing, transforming, and A/D conversion can be realized with the SI technique. The preprocessor block compensates for the dark

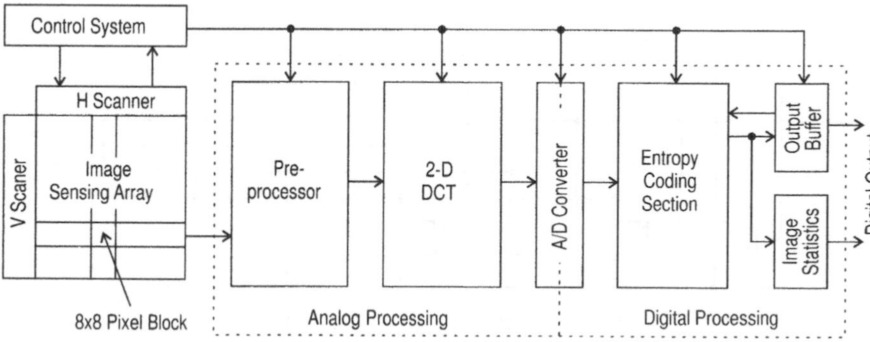

Figure 7.1 One-chip digital video camera architecture.

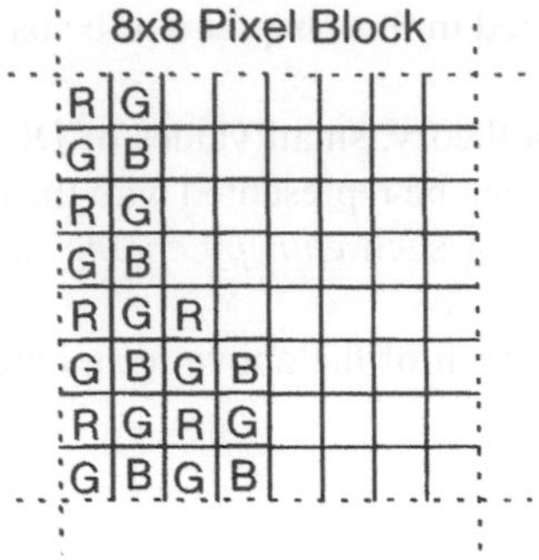

Figure 7.2 Pixel block covered by red, green, and blue (RGB) filters, following the Bayer pattern.

current of photodiodes and produces balanced signals for further processing. A regulated-gain amplifier (RGA) is used for gain control and white balancing. The RGA drives the color interpolation and space conversion circuit (SCC). The signal obtained in luminance and chrominance space is compressed with the use of a DCT processor. The output block of the DCT section performs two operations: the inversion of one of the signals in each balanced pair and the addition of these two signals. The digital part of the chip produces entropy encoded video data that is obtained at the output in the form of a digital string.

7.2.1 Current-Mode Regulated Gain Amplifier (RGA)

R, G, and B optical filters, which cover the pixels, introduce unequal attenuations. In order to compensate for these different attenuations, an amplifier with digitally regulated gain is used. Such an amplifier, based on the current mirror structure, is shown in Figure 7.3. We see that with the use of transistors with scaling factors of

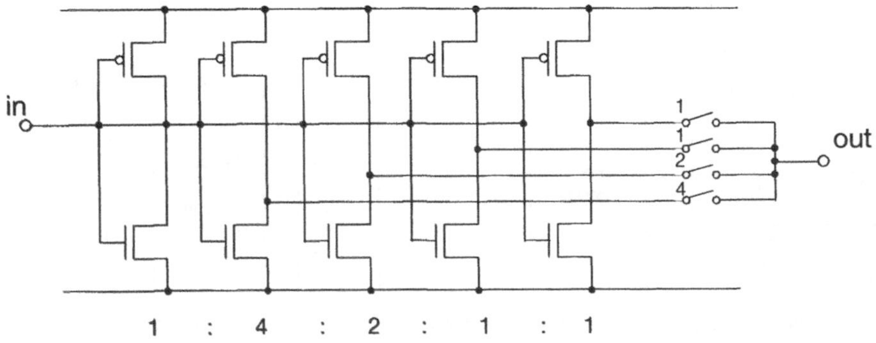

Figure 7.3 Regulated gain amplifier (RGA).

channel widths equal to 1, 2, and 4, eight values of the gain can be obtained. The circuit also allows amplification of the pixel signals during low-light operation. An interpolation circuit follows the RGA to calculate color components in the pixels omitted with respect to the Bayer pattern.

7.2.2 Color Space Conversion Circuit (CSCC)

The current-mode color space converter transforms the red-green-blue (*RGB*) space to the YC_RC_B space. The circuit is designed on the basis of equations (7.1). It is easy to implement such operations on matrices in the SI technique with the use of current mirrors. In order to describe this approach in detail, let us consider the matrix equation in the form

$$\begin{bmatrix} y_1 \\ y_2 \end{bmatrix} = \frac{1}{k}\begin{bmatrix} a_{11} & a_{12} \\ a_{21} & a_{22} \end{bmatrix} \cdot \begin{bmatrix} x_1 \\ x_2 \end{bmatrix} \tag{7.2}$$

The schematic diagram of the implementation of this matrix relation is given in Figure 7.4. For minimization of parasitic effects, the balanced structure is used. The matrix elements are realized as the current mirrors with scaling factors equal to the values of the matrix elements. The currents at the outputs of mirrors are equal to appropriate components in the matrix equation (7.2). The outputs of current mirrors are connected to each other in order to add the current signals and to realize the whole matrix equation. In Figure 7.4 we show these connections, assuming that the matrix elements a_{ij}, $i, j = 1,2$ are positive. Because of the balanced structure, it is easy to obtain connections for negative values of elements, too. The block diagrams of the SI cells realizing relations (7.2) and (7.1) are presented in Figure 7.5 and Figure 7.6, respectively.

7.2.3 Two-Dimensional DCT Processor Implementation

Relations (1.36) shows that the two-dimensional (2-D) discrete cosine transform (DCT) can be realized by two one-dimensional (1-D) ones. The matrix X in (1.36) denotes one input 8×8 block, and its transposition X' in the relation $Z = X'C'$ means that it is read column by column. Matrix Z, containing intermediate results, is also obtained with the use of 1-D DCT, and saved in a memory array. Transposition of this matrix in relations (1.36) means that the elements of Z obtained successively for the current block are memorized in row cells and are read from column cells of the memory for the previous block. The intermediate results are processed in the same way as the input matrix X, giving the output signal matrix Y for the entropy coding section, as shown in Figure 7.1.

The implementation of a 1-D DCT block is presented in Figure 7.7. It is composed of 16 current mirrors with the scaling factors (a, b, c, d, e, f) corresponding to the elements of the matrix C' (1.35). The outputs of current mirrors are connected to each other, as shown in Figure 7.4, in order to add the current signals and to realize matrix relations (1.36).

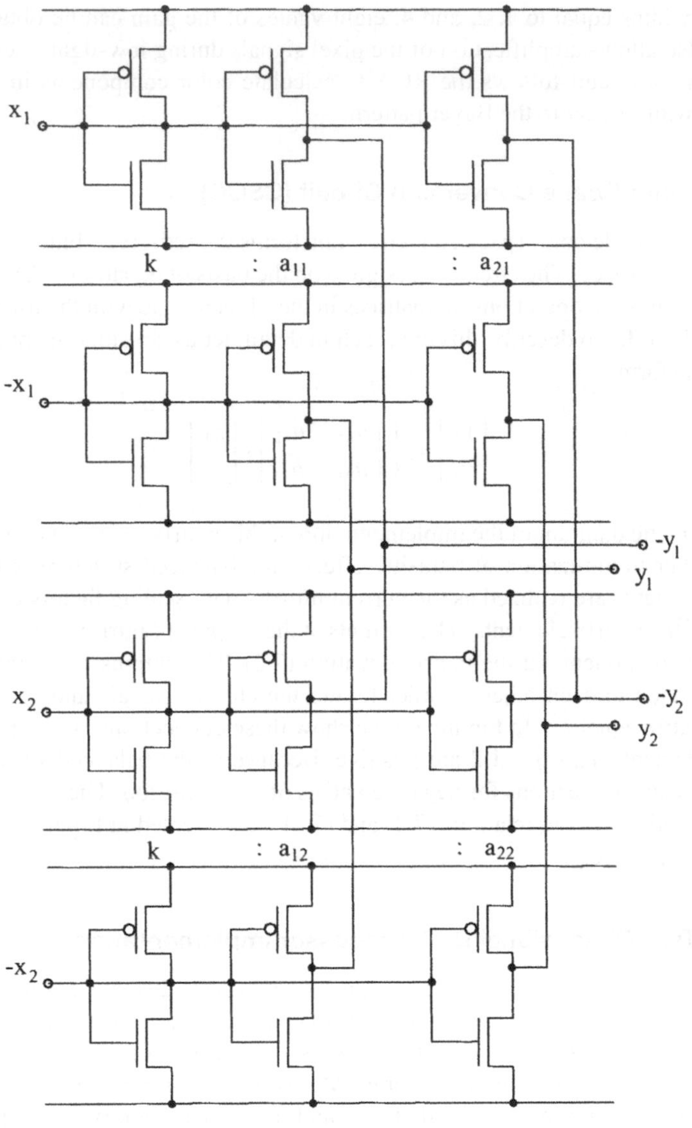

Figure 7.4 Schematic diagram of a matrix relation between signals x and y in an SI balanced structure, implemented with the use of multioutput current mirrors.

The memory array implementation with the use of delay elements is shown in Figure 7.8. The memory is controlled by the sixteen-phase clock, which is synchronized with a horizontal scanner. An example of a clock that can be used in such applications is presented in Figure 4.25. The schematic diagram of a delay element is presented in Figure 7.9, [24]. The delay cell is realized in the balanced structure, which is very useful for compensation of parasitic effects like dc offset, clock feed-

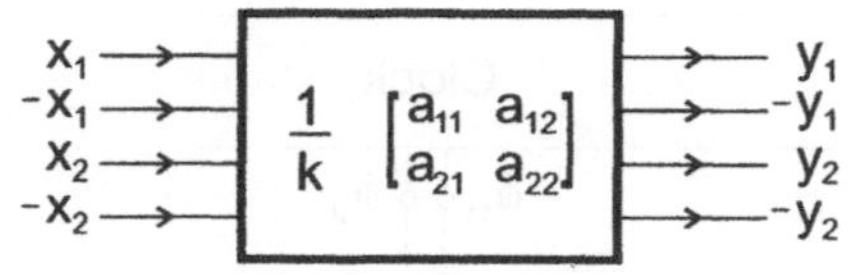

Figure 7.5 Block diagram of a matrix relation between signals x and y.

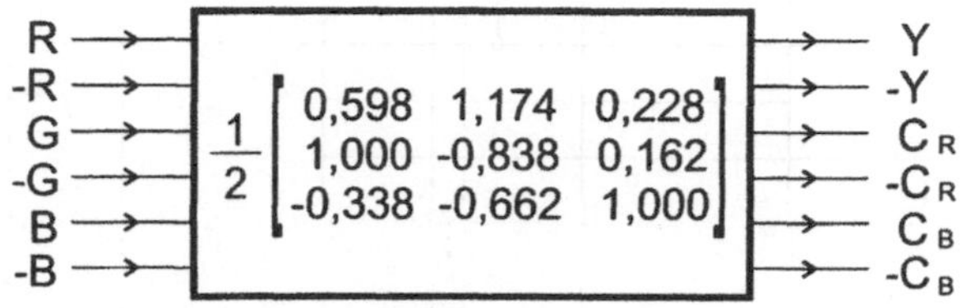

Figure 7.6 Block diagram of the color space conversion circuit (CSCC).

through, and charge injection. Let us note that the signal is sampled in each half of the clock period. Hence, the sampling frequency is twice as great as the clock frequency. This property makes it possibile to write the intermediate results for the current image block in one phase of the clock period and to read the intermediate results for the previous block in the other phase of the clock period. The 2-D DCT processor is depicted in Figure 7.10.

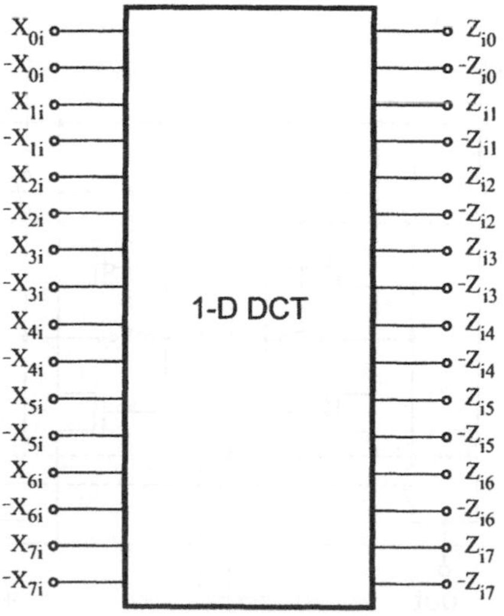

Figure 7.7 Current mode implementation of 1-D DCT.

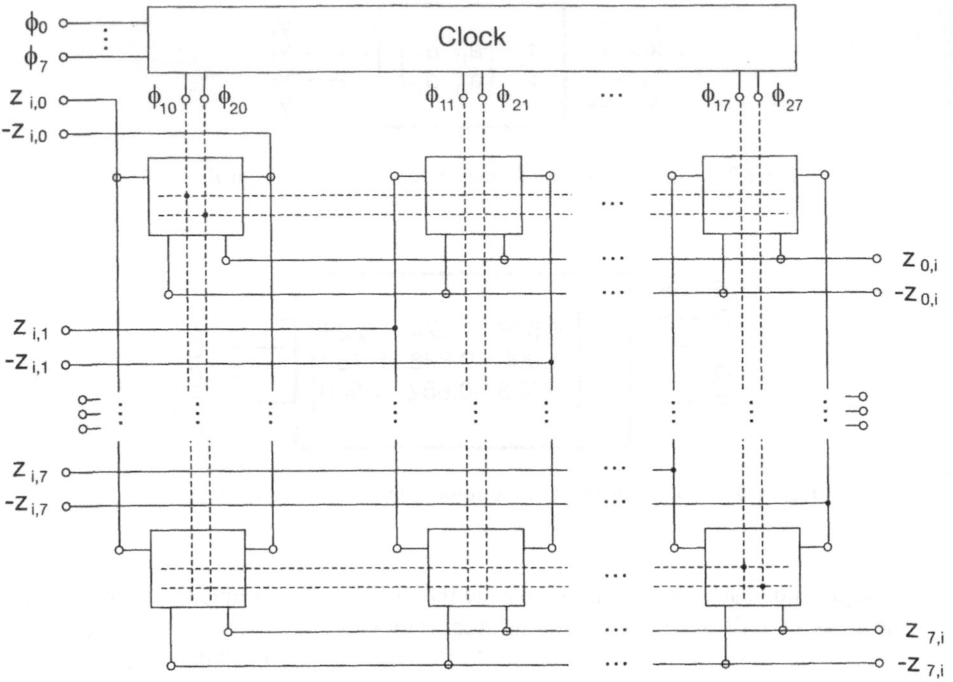

Figure 7.8 SI realization of the memory array composed of delay elements.

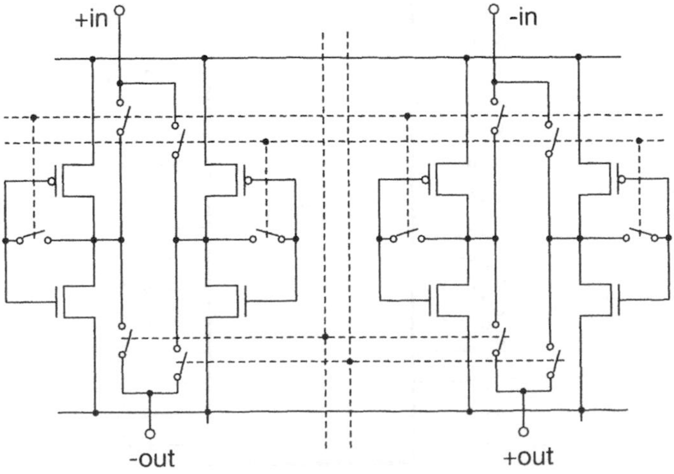

Figure 7.9 Schematic diagram of a delay element.

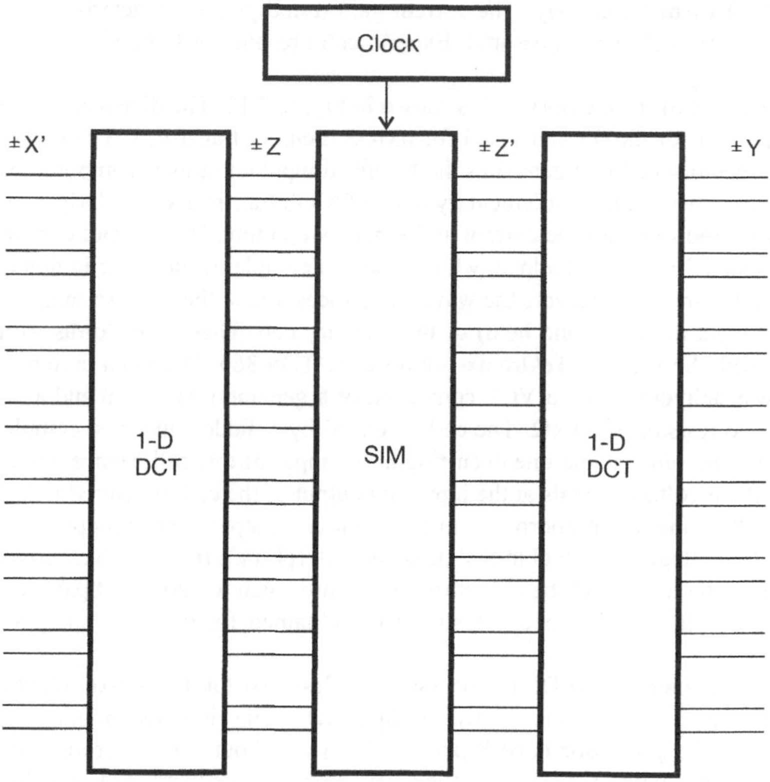

Figure 7.10 2-D DCT processor composed of two 1-D DCT cells and the SI memory macrocell.

7.2.4 Entropy Switched-Current Coding Section

The input stage of the entropy coding section is an analog-to-digital converter. Converters based on delta-sigma modulators and described in Section 6.1 seem to be the most useful for switched-current implementation. The basic cells of such modulators are integrators and were discussed in the first part of the book. The bilinear integrator, realized in the classic structure, is presented in Figure 5.8. We can observe from (5.22) that parasitic effects, especially the gain error δ, can be better compensated for in the classic integrator than in the one composed of Euler integrators. The large range of damping coefficient regulation is another useful property of the classic integrator. Simulations show that the sampling frequency necessary for video applications is achievable in the converter based on the classic integrator.

7.2.5 Results of Experiments

In this section, we will show experimental verification of the considered cells. The prototype chip containing current mirrors, delay cells, and integrators was designed in AMS 0.8 μm technology. The current gain α and phase characteristic ϕ of the current mirror cell were measured. Experimental results for these circuits are presented in [25].

The layout of the memory cell is shown in Figure 7.11. The dimensions of channels are 56/4 for nMOS and 120/4 for pMOS memory transistors. The input stage for the memory cell that generates the balanced input currents is composed of current mirrors based on complementary pairs of MOS transistors. Similarly, a current mirror is used to invert the current at the negative output. Both output currents are added and delivered to the load, which is also a complementary pair in diode connection. Figure 7.12 presents the waveforms measured at the input (triangular) and at the output (sampled and held) of the memory cell. These waveforms were obtained with the use of a Tektronix oscilloscope TDS 360. The input current of the cells was delivered from a VCT composed of a generator MS 9160 and a serially connected resistor of 20 kΩ. The cell is loaded by a diode-connected complementary pair, the same as the one occurring at the input of this cell. Hence, the amplitudes of the voltage signals at the input and output of the cell, measured with a digital oscilloscope, are proportional to the input and output current amplitudes. The generator voltage was equal to 8 V peak-to-peak (pk-pk). It means that current amplitudes at the input and the output of the cell are equal to about 400 μA pk-pk for the resistor 20 kΩ. The above results were obtained for the low supply voltage ±1.35 V.

An experimental 2-D DCT processor was designed for 4 × 4 pixel blocks. The layout of the 2-D DCT is presented in Figure 7.13. The chip area, in 0, 8 μm technology, of the processor is 61.8 μm × 1064 μm (about 1 mm^2) from which the memory array takes 523.2 μm × 1064 μm. Experiments were made for the same values of pixels in the rows of 4 × 4 input block. The waveform at the first output is shown in Figure 7.14. The values at the second, third, and fourth outputs are equal

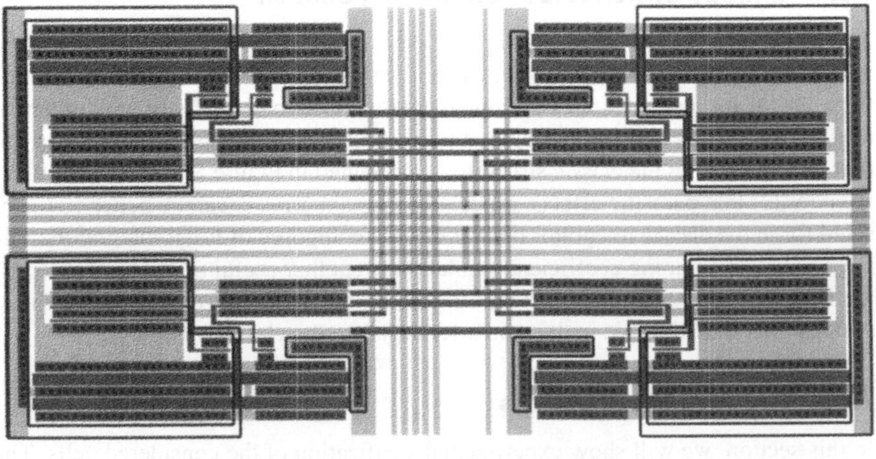

Figure 7.11 Layout of a delay element.

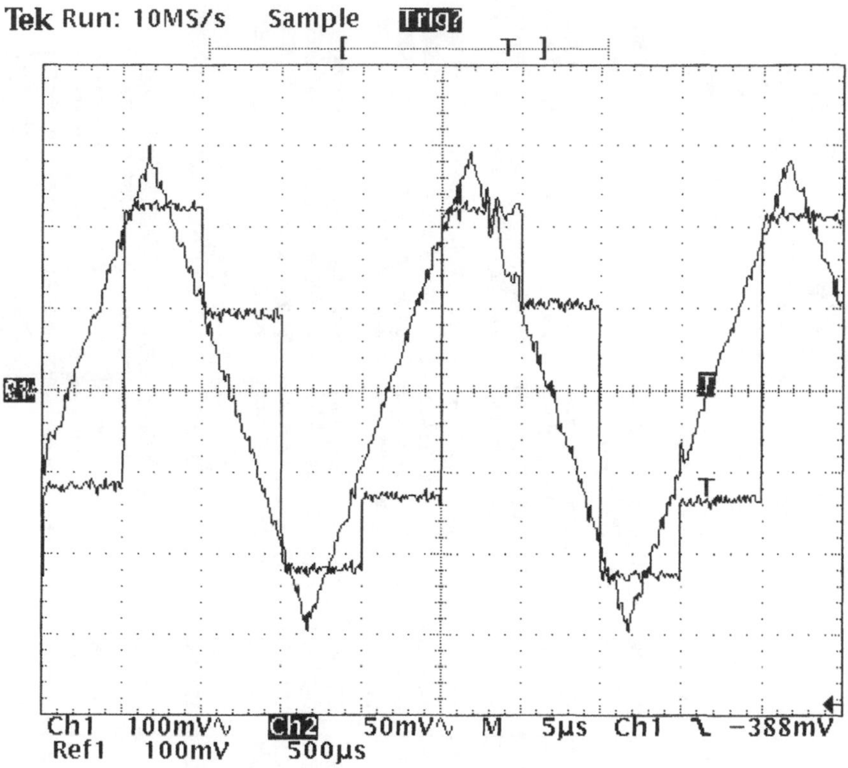

Figure 7.12 Sampled and held output waveform of the memory cell.

to zero. The voltages are measured at the input stage and at the output load in the same manner as for a single memory cell. The exact values of matrices X and Y for the waveform in Figure 7.14 are

$$X = \begin{bmatrix} 50 & 50 & 50 & 50 \\ 0 & 0 & 0 & 0 \\ 50 & 50 & 50 & 50 \\ 50 & 50 & 50 & 50 \end{bmatrix} \quad Y = \begin{bmatrix} 150 & 0 & 0 & 0 \\ -27 & 0 & 0 & 0 \\ 50 & 0 & 0 & 0 \\ 66 & 0 & 0 & 0 \end{bmatrix} \tag{7.3}$$

For the measured values of Y shown in Figure 7.14, the inverse 2-D DCT was calculated and compared to X in (7.3). Obtained $PSNR = 38.56$ dB. Other experiments gave similar results. Let us note that in the experiment small voltages were measured. Hence, it can be expected that using current mode A/D converter at the output of DCT processor will give much better results. This converter, based on the Δ/Σ modulator, is described in Section 6.1.

It can be seen from Figure 7.14 that the sampling frequency is 1 MHz. Increasing the frequency up to 4 MHz does not significantly change the results. The results

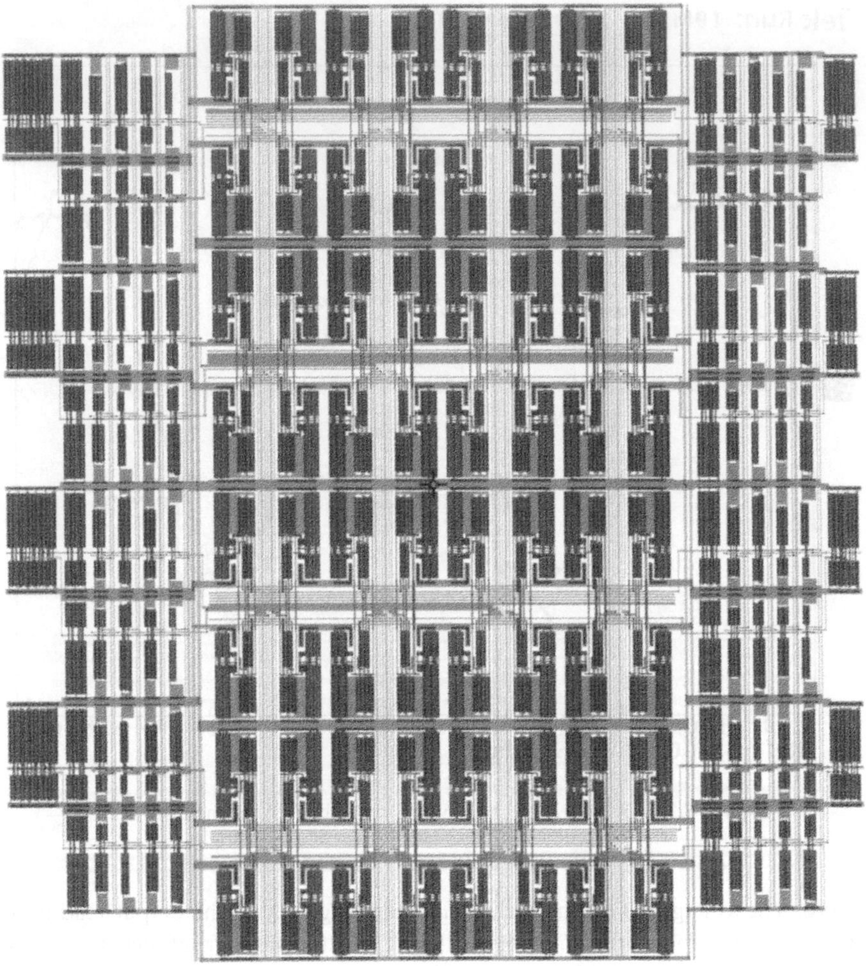

Figure 7.13 Layout of the 2-D DCT processor.

show that it is possible to decrease dimensions of transistors in order to reduce the power consumption and to increase the speed.

7.3 2-D SIGNAL PROCESSING WITH THE USE OF A MULTIPORT NETWORK

In this section, we will show that the problem of the design of a 2-D system for real-time processing can be reduced to the problem of multiport network synthesis. In order to prove this, we will use a hybrid representation of the transfer function in the form

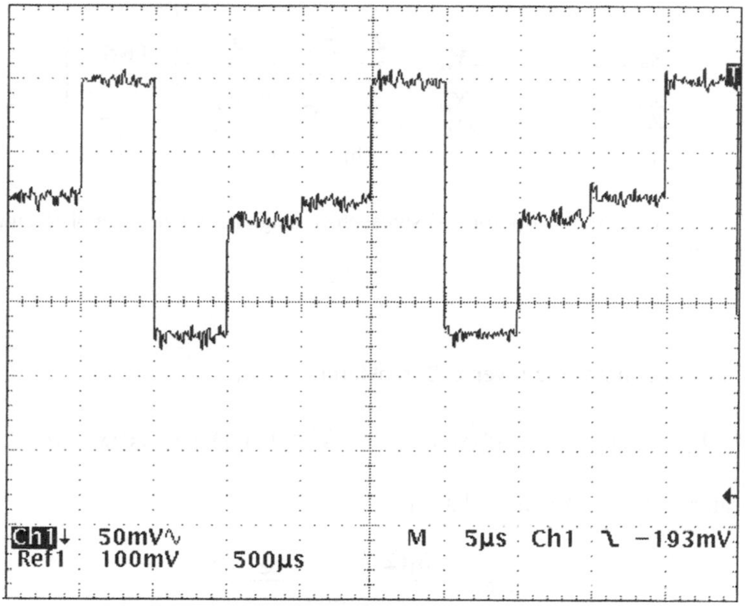

Figure 7.14 Experimental output waveform of the 2-D DCT processor.

$$\sum_n H^{mn}x^n = \sum_n \frac{S_1 P_h^{mn} Z_2'}{S_1 Q_h Z_2'} x^n(s_1, z_2) = y^m(s_1, z_2) \tag{7.4}$$

(see Figure 1.4). We can write this relation as follows

$$\sum_n S_1 P_h^{mn} Z_2' x^n(s_1, z_2) = S_1 Q_h Z_2' y^m(s_1, z_2) \tag{7.5}$$

We decompose the matrix Q_h into submatrices Q_c and $-Q_r$ so that

$$Q_h = [-Q_r \;\vdots\; Q_c] \tag{7.6}$$

where the matrix Q_c is equal to the last column of Q_h and Q_r is the residue of the matrix Q_h taken with the opposite sign. Then the multiplication of the vector Z_2' by the functions x^n and y^m in (7.5) yields

$$\sum_n S_1 P_h^{mn} X^n = -S_1 Q_r Y^m + S_1 Q_c Y_i^m \tag{7.7}$$

where

$$X^n = [X_{i-k}^n \cdots X_{i-1}^n \, X_i^n]', \; Y^m = [Y_{i-k}^m \cdots Y_{i-1}^m]' \tag{7.8}$$

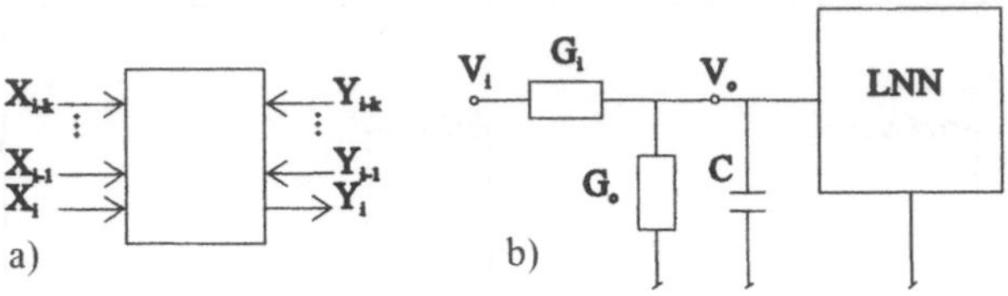

Figure 7.15 Multiport network for 2-D signal processing (a), and the symbol of a multiport prototype network (b).

and the index i denotes the inverse Z transform

$$Y_i^m = Y_i^m(s_1) = Z^{-1}\{y^m(s_1, z_2)\}, \qquad X_i^n = X_i^n(s_1) = Z^{-1}\{x^n(s_1, z_2)\} \tag{7.9}$$

We can rewrite (7.7) as the relation

$$Y_i^m(s_1) = \frac{S_1(\Sigma_n P_h^{mn} X^n + Q_r Y^m)}{S_1 Q_c} \tag{7.10},$$

which describes a multiport network (Figure 7.15a) excited by the signals $X_{i-k}^n, \ldots,$ $X_i^n, Y_{i-k}^m, \ldots, Y_{i-1}^m$ and with the responses Y_i^m at the output ports.

Let us note that the polynomial S_1Q_c in (7.10) is equal to $S_1Q_hZ_2'$ for $z_2^{-1} = 0$, or to S_1QS_2' for $s_2 = 1$. Hence, if we assume that the denominator of the transfer functions H^{mn} is a 2-D Hurwitz polynomial, i.e., it has no zeros in the domain $res_1 \geq 0$, $res_2 \geq 0$, then the denominator S_1Q_c is also a Hurwitz polynomial.

7.4 PROTOTYPE CIRCUIT OF A MULTIPORT 2-D FILTER

In this section, we describe the design method of the multiport 2-D network presented in Figure 7.15a, on the basis of a multiport lossless prototype network. The method is similar to the one presented in Chapter 6 in the case of low-sensitivity design strategy for 1-D ladder filters. The possibility of using a multiport lossless network is implied by the denominator in (7.10), which is a Hurwitz polynomial. However, the prototype network cannot be simply a reactance network; in general, it must be a lossless nonreciprocal network. It can be a gyrator–capacitor network [20]. A model of this prototype network, containing capacitors and a lossless noninertial network (LNN) composed of grounded gyrators, is presented in Figure 7.15b. There exist many synthesis methods for designing lossless multiport networks terminated in resistors [44]. If such a network is terminated in impedances, then the synthesis problem is solvable with broadband matching theory [8].

We will synthesize the network shown in Figure 7.15b with the coefficient matching method. In order to match $Y_i^m(s_1)$ in (7.10) with the output voltage $V_o^m(s_1)$ of this prototype network, it is assumed that the vector V_i of excitations V_i^l, $l = 1, \ldots, L$, where $L \geq M$, is defined as a linear combination of the signals X^n and Y^m:

$$V_i = [V_i^1 \cdots V_i^L]' = \sum_{n=1}^{N} B_n^x X^n + \sum_{j=1}^{M} B_j^y Y^j \tag{7.11}$$

where B_n^x and B_m^y are coefficient matrices. Hence, the relation describing the output voltage $V_o{}^m$ is as follows:

$$V_o^m(s_1) = \frac{S_1 M^m V_i}{S_1 M_d} = \frac{S_1(M^m \sum_{n=1}^{N} B_n^x X^n + M^m \sum_{j=1}^{M} B_j^y Y^j)}{S_1 M_d} \qquad m = 1, \ldots, M, \tag{7.12}$$

where elements of the matrices M^m and M_d are functions of the prototype network parameters.

Matching the denominators and numerators in (7.10) and (7.12) yields

$$\begin{aligned} M_d &= Q_c \\ M^m B_n^x &= P_h^{mn} \\ M^m B_j^y &= Q_r, j = m \\ M^m B_j^y &= 0, j \neq m \qquad m = 1, \ldots, M, n = 1, \ldots, N \end{aligned} \tag{7.13}$$

The above matrix relations denote a set of nonlinear algebraic equations. In these equations, the unknown variables denote the parameters of prototype network elements and elements of the matrices B^x and B^y. We can solve this set of equations in many ways. The tools based on Fletcher–Powell optimization algorithm are used in a system presented in Chapter 8.

7.5 SC AND SI TWO-DIMENSIONAL FILTERS

We will illustrate the method of 2-D filter design by an example of a detail-emphasis image filter. The filter, described in [56], has the transfer function in the form of (1.45) with the matrices

$$\hat{A} = 2.5 \begin{bmatrix} -0.1557553 \cdot 10^{-1} & 0.468344 \cdot 10^{-1} & -0.399411 \cdot 10^{-2} \\ 0.4951426 \cdot 10^{-1} & -0.2103131 \cdot 10^{-1} & -0.2237115 \\ 0.411281 \cdot 10^{-2} & -0.2336235 & 0.707513 \end{bmatrix} \tag{7.14}$$

and

$$\hat{B} = \begin{bmatrix} 1 & 0.223576 & 0.7149619 \cdot 10^{-1} \\ 0.22108698 & 0.1544512 & 0.1057191 \\ 0.9173054 \cdot 10^{-1} & 0.1020029 & 0.15625 \end{bmatrix} \tag{7.15}$$

The sign $\hat{}$, introduced in the first chapter, denotes a matrix transposed with respect to both diagonals.

Figure 7.16 shows an example of an image before and after processing with the use of the filter. The matrices P_h and Q_h in (7.4) are, from equations (1.57), in the form

Figure 7.16 Example of an image before (top) and after (bottom) processing with the use of the detail-emphasis filter.

$$P_h = T_2' A, \qquad Q_h = T_2' B \tag{7.16}$$

where the matrices A and B correspond to variables z_1 and z_2 in the ordering given by (1.44) and T_2 is described by (1.49). The prime sign denotes a transposed matrix. The matrix Q_h is decomposed into Q_c and Q_r according to (7.6).

The prototype gyrator–capacitor circuit is shown in Figure 7.17. In order to calculate the values of parameters of the prototype circuit we must solve the equations (7.13), which for this example have the form

$$\begin{aligned} M_d &= Q_c \\ MB^x &= P_h \\ MB^y &= Q_r \end{aligned} \tag{7.17}$$

On the basis of the symbolic analysis of the prototype circuit, the elements of matrices M_d and M are determined as functions of circuit parameters. Computer tools for symbolic analysis are described in the next chapter. With the use of the Fletcher–Powell optimization method, the first set of algebraic equations $M_d = Q_c$ is solved for the initial point $x_0 = 0.65$, giving the following values of the prototype circuit parameters:

$$\begin{aligned} &C_1 = 0.635042, \quad C_2 = 0.454336, \quad C_3 = 0.53273 \\ &g_1 = 0.633176, \quad g_2 = 0.766516, \quad g_3 = 0.61323 \\ &G_1 = 0.557962, \quad G_2 = 0.5720097, \quad G_3 = 0.622655 \\ &G_4 = 0.446344 \end{aligned} \tag{7.18}$$

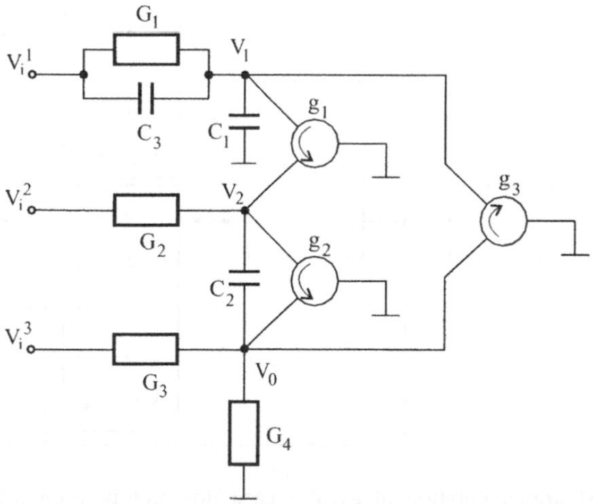

Figure 7.17 Gyrator–capacitor prototype circuit for the detail-emphasis filter.

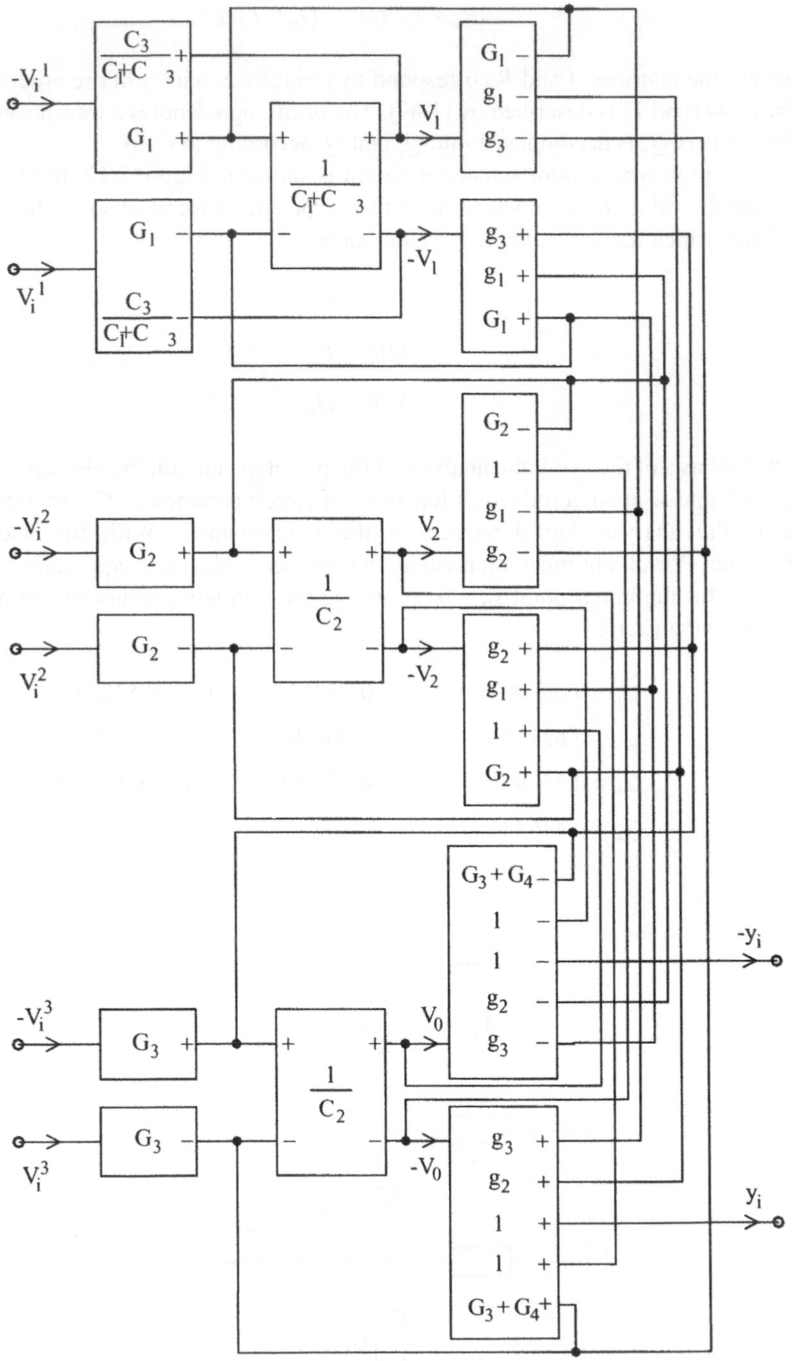

Figure 7.18 SI implementation of gyrator–capacitor prototype circuit for a picture enhancement 2-D filter.

For these values of parameters, the elements of matrix M are calculated. From the last two matrix equations of (7.17) we obtain

$$B^x = M^{-1}P_h = \begin{bmatrix} 31.650229 & -21.511419 & 5.845858 \\ -75.03833 & 27.653978 & -1.794539 \\ 75.49955 & -26.348482 & 1.101939 \end{bmatrix} \tag{7.19}$$

and

$$B^y = M^{-1}Q_r = \begin{bmatrix} -8.294074 & -6.824212 \\ 7.300033 & 4.006445 \\ -6.955436 & -4.099399 \end{bmatrix} \tag{7.20}$$

All the above calculations and matrix operations are performed with the use of computer tools that will be described in the next chapter. An SC implementation of the detail-emphasis filter is presented in [21]. In order to obtain an SI implementation, we can write the node voltage equations of the gyrator–capacitor prototype circuit in the following form:

$$V_1 = \frac{1}{s(C_1 + C_3)}[-G_1V_1 - g_1V_2 + g_3V_o + G_1V_i^1] + \frac{C_3}{C_1 + C_3}V_i^1$$

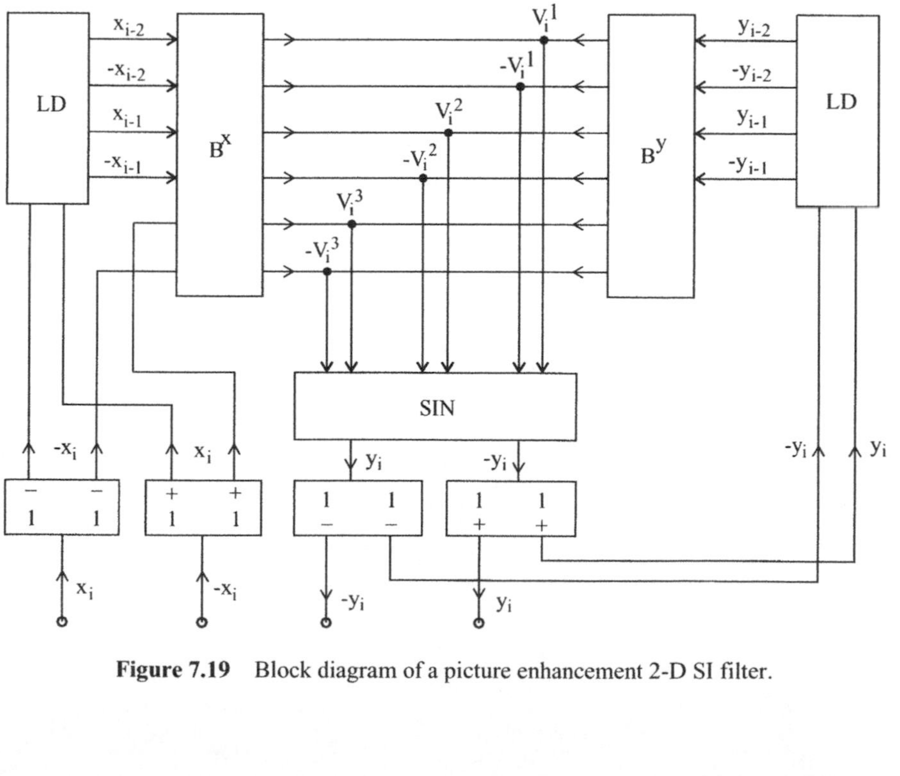

Figure 7.19 Block diagram of a picture enhancement 2-D SI filter.

$$V_2 = \frac{1}{sC_2}[g_1 V_1 - G_2 V_2 - g_2 V_o + G_2 V_i^2] + V_o$$

$$V_o = \frac{1}{sC_2}[-g_3 V_1 + g_2 V_2 - (G_3 + G_4)V_o + G_3 V_i^3] + V_2 \tag{7.21}$$

The SI counterpart network (SIN), depicted in Figure 7.18 and composed of integrator and current mirror cells, can be obtained in the same way as the SI filter in Figure 6.27 which was obtained on the basis of equations (6.75).

The obtained SI 2-D filter is shown in Figure 7.19. The basic macrocell is the SI network (SIN) shown in Figure 7.18. The cells B_x and B_y perform multiplications by matrices that appear in relation (7.11). The implementation of this operation is shown in Figure 7.4. We realize the delay lines (LD) with the use of the memory cell given in Figure 7.9. Each row of the memory array shown in Figure 7.8 is an example of a delay line.

7.6 PROBLEMS

1. Transform the function given in the analog domain by (1.61) into the hybrid domain $H'(s_1, z_2)$ and determine the matrices P_h, Q_r, and Q_c in (7.10).

8

Silicon Compilers

Software systems that can automatically generate integrated circuits are called silicon compilers. They contain silicon assembly tools that generate the layout of a circuit on silicon. Professional systems also perform relatively hardware-independent behavioral synthesis on the basis of specifications given by the designer.

There is a number of highly effective systems for design of digital circuits, e.g., Cadence, Compass, Mentor Graphics, Synopsis, and Lager. However, tools that can be used in mixed-signal integrated circuit design are being intensively developed. This chapter presents briefly the basic ideas that are necessary for the use and development of such systems. Topological methods of analysis that are very helpful in the design of filters at the level of behavioral synthesis are discussed in more detail.

8.1 OBJECT-ORIENTED DATA BASES

In the first part of the book, we described different kinds of cells that are the basic units of integrated circuits. A cell can consist of other cells. For example, an amplifier is composed of two or more stages, a full-adder contains simpler digital gates, etc. The design process is hierarchical, which means that the designed cell, called the *macrocell*, is composed of cells that can contain other *subcells*, etc. Typical cells are delivered from a foundry in a *library* of cells. In the design, each reference to a given subcell is called an *instance* of that subcell. We can say that an instance is a particular placement of a subcell in a cell or that a placement is a process of creating instances. The cells that are used in the placement are often called the *leafcells*. In a data base of a silicon compiler used in the design, each cell has a counterpart called an *aspect* or a *view*. Aspects (views) of cells depend on the design level. We distinguish the following views:

- The *symbolic view,* which is used at a high level of abstraction. At this level, the cell can be represented by different graphical or mathematical symbols.
- The *schematic view* presents the subcells that the cell consists of. It also shows how the subcells are connected.
- The *simulation view* is a description of the cell that allows one to use a chosen simulator.
- The *physical view* is simply the layout of the cell.
- The *phantom view* shows the size and shape of the cell on a chip.

The symbolic view is usually used in behavioral synthesis. The design process is hierarchical, which means that we work at a chosen level of this hierarchy. At this level, the subcells can be either not yet designed or taken from a library and represented by their phantom views. The cell edited at a given level of the hierarchy is called the *composition cell* or the *facet* of the view. In a data base, the cells have names, and in order to distinguish between different views (aspects), suffixes are added to the names. For example, let *3filter* be the name of a third-order filter cell. Then *3filter[ly]* denotes the physical aspect of this cell, *3filter[cir]* denotes the simulation view in the SPICE simulator, and *3filter[cp]* denotes the composition cell.

8.2 INTERFACING WITH SILICON COMPILERS

Silicon compilers allow the designer to enter a circuit as a schematic (schematic entry). The symbols of cells like those shown in the first part of this book are entered as components in the schematic. Graphical representation is useful for integrated circuits composed of a relatively small number of cells.

For complex designs that use parameterization, there are different kinds of textual description. Verilog and VHDL are popular languages that are accepted by many silicon compilers for the design of digital integrated circuits. VHDL is an acronym for very high speed integrated circuit (V) and hardware description language (HDL). Computer systems for integrated circuit design can also have their own specific languages. They use the syntax of higher-level languages, for example the C-like or LISP-like syntaxes, which are advantageous to a user. Professional compilers can automatically generate the layout of digital circuits from textual descriptions. For analog circuits, the textual representation and tools for the design process are intensively developed [10, 28]. VHDL–AMS (analog and mixed signal) is a new IEEE–VHDL 1076.1 standard obtained as an extension of the mature VHDL environment. However, the possibilities of VHDL–AMS in the design of analog cells are not yet as wide as in the design of digital cells.

The basic unit or system on which we operate and which must be declared in VHDL is called an *entity*. The already created entities are available in libraries and allow the synthesis of digital circuits using different technologies. The parameters of an entity are declared with the use of a *generic* whereas the connections are declared with the use of a *port*.

In order to describe data processing inside an entity, *architecture* is used that has syntax like the function syntax in programming languages. Each entity can possess more than one architecture. The parameters, as in programming languages, can be integer, real, and array variables. However, in VHDL there is a special variable called a *signal*. Signals are variables that contain information on their current and previous values. Hence, apart from multiply, sign, add, shift, relational, and logical operators there are expressions assigned to signals. These expressions are delays like *after*, *inertial*, and *transport*. Another important keyword in VHDL is *process*. It is used in order to describe a list of sequential operations that can be performed in the system. Such operations are typical for finite-state machines.

As we mentioned, VHDL–AMS is an extended version of VHDL. The most important keyword of this extension is *quantity*. Analog circuits are described by a set of differential equations, which can be written in the form

$$F(x, x', t) = 0 \tag{8.1}$$

where x is the vector of the unknown functions and the prime denotes derivatives of functions with respect to time t. Such equations are solved in an algorithmic manner and the *quantity* is used to introduce unknown functions in the description of analog circuits. There are two kinds of quantities: *across quantities*, representing voltages; and *through quantities*, representing currents in graph branches of a circuit. Quantities may also represent mechanical, thermal, or other physical *nature* phenomena.

8.3 ASSEMBLY

Silicon assembly involves procedures that directly lead to layout generation. Assembly is a time-consuming process. Hence, it is very important to automate it. The starting point in this process is a schematic entry or a textual description of the circuit structure. This section presents the basic ideas that each professional silicon compiler must master.

8.3.1 Tiling and Floor Planning

In a process called tiling, a silicon compiler assembles the layout of a macrocell by abutting instances of leaf cells. This task is relatively simple in the case of digital circuits, for which a macrocell is usually implemented as a regular array of subcells. The interconnection between leaf cells is made through abutment. The instances of the subcells can be overlapped, rotated, mirrored, and then abutted to form a macrocell.

Another situation occurs when the cell is composed of subcells that are of a completely different kind. Such components can be, for example, obtained through a tiling process. This task, realized by placement of subcells, channel definition, and routing, is called floor planning. For macrocell-based design, the floor planning tools can work interactively. If the system allows one to create an automatic floor plan, then this floor plan can be used as an initial solution. This initial solution is then interactively improved. In order to interconnect the macrocells, channels must be created between them. In Figure 8.1 a floor plan of macrocell instances is shown. Figure 8.1 also shows the channels defined in the floor plan and the wires created in the routing process.

As seen in Figure 8.1, routing is the process of creating wires to connect instances. Routing usually refers to automatic routing. If it is manual, it is called wiring.

8.3.2 Datapath Generation

Multibit digital circuits often require more effective tools than tiling and floor-planning. Some silicon compilers offer a bit-slice datapath strategy, illustrated in Figure 8.2.

The macrocell in Figure 8.2 is composed of slices, each of them corresponding to one bit. Figure 8.2 shows the less significant bit (LSB) slice and the most signifi-

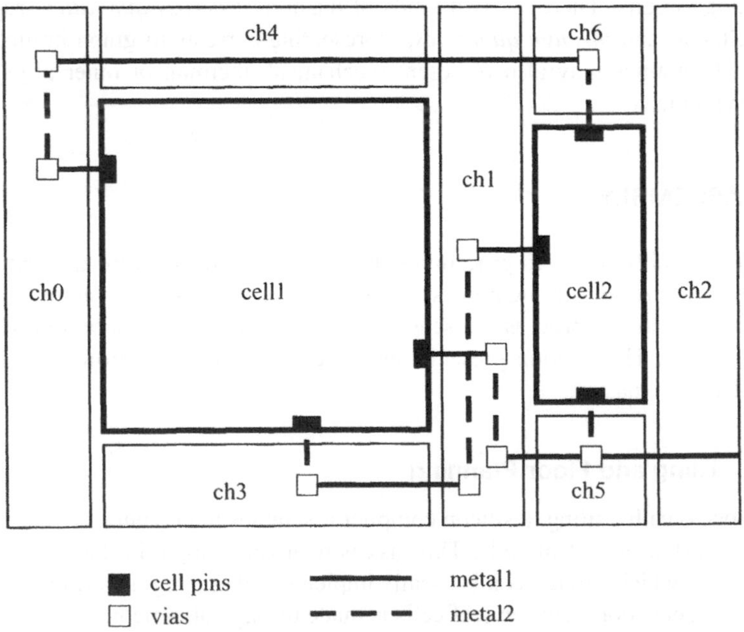

Figure 8.1 Floor plan containing instances of macrocells, channels, and wires.

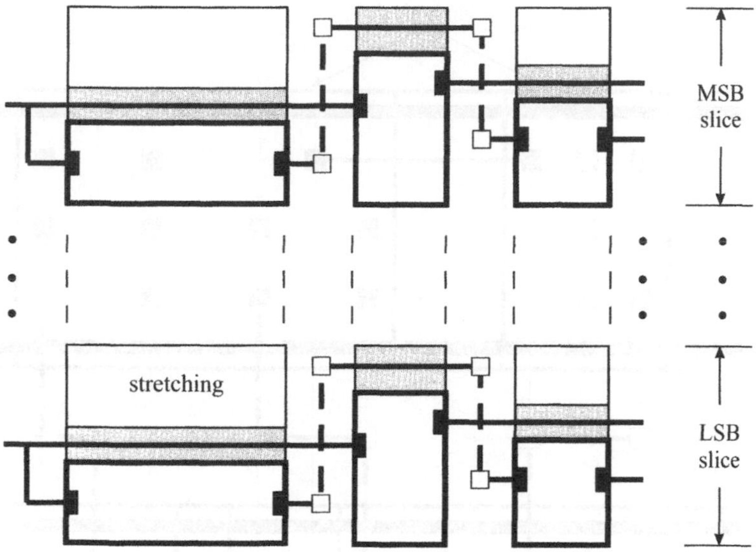

Figure 8.2 Example of a bit-slice datapath strategy.

cant bit (MSB) slice. Each instance of the cell shown in Figure 8.2 consists of a leaf cell and two optional cells, the first, (black) is used for feedthrough connections and the second (white) is called a stretching cell. In this strategy, data buses can be routed not only in the data channels but in these optional cells, too. The strategy makes the cells porous and stretchable, shortens connections in a circuit, and results in reduction of the total chip area.

8.3.3 Area Router

The interconnections described above, based on channel definition, are mainly used for two-metal routing. Figure 8.1 and Figure 8.2 illustrate this method of routing, which is called maze-style routing. For three or more metal layers, the method of routing is based on area routers. In this method, the cells are placed in a structured layout, usually in rows, and pins are located not on the perimeter but on the whole area of the cell. It is assumed in this method that power pins are on metal 1 and have horizontal access. Metal 1 is used to realize connections inside cells, which limits its application for interconnections. Signal pins have vertical access. Therefore it is advantageous to use metal 2 as a vertical and metal 3 as a horizontal routing layer.

Standard cells utilize rows in less than 100% of cases. Filler cells are introduced in spaces between standard cells and cap cells should be placed on each end of a row. Routing can be performed in spaces between rows, through the filler cells, and over the standard cells. Figure 8.3 shows cells placed in rows and covered by a routing grid. The pins of the cells are exactly on the grid points. Black boxes in these

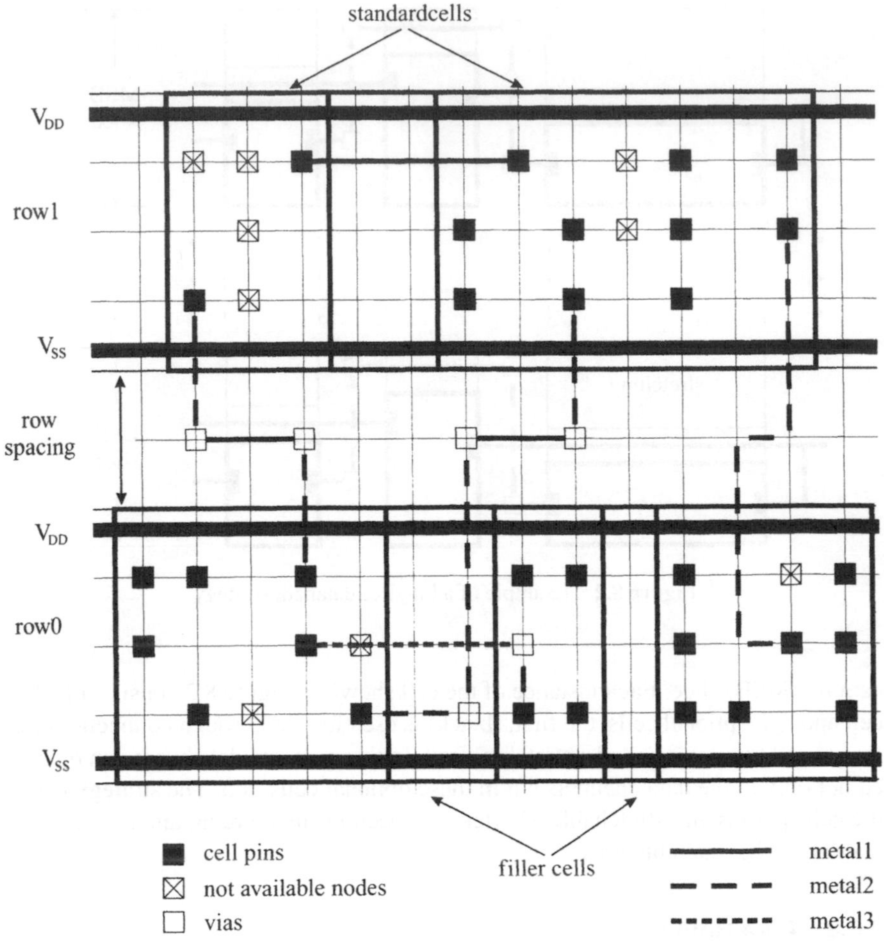

Figure 8.3 Routing based on area router.

points denote contacts, whereas the boxes with a cross inside are not available to metal 1. Connections are routed along the lines of the grid.

8.3.4 Pad Routing

The core part of a chip is surrounded by a bonding pad ring. Libraries contain input pads, which buffer and protect the inputs of the core, and output pads, which provide drive capability of outputs of the core. Pads are collected into groups, as shown in Figure 8.4.

Corner pieces and space pads are added when the core is too large to close the pad ring. Additional space pads can be added later if the routing space is found to be too small for the nets connecting the core part and the pads.

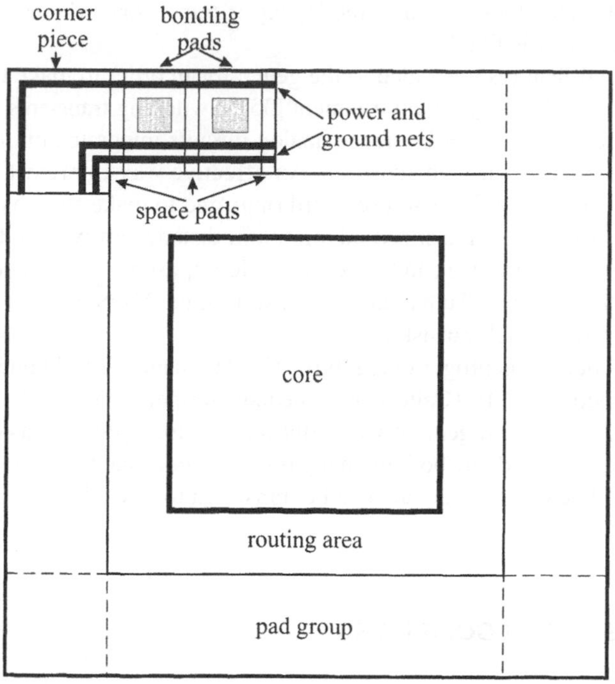

Figure 8.4 Bonding pad ring around the core part of the chip.

8.4 LAYOUT AUTOMATION OF ANALOG CELLS

There exist libraries that are offered by silicon foundries as technological files and which contain digital elements. Pads are included in the libraries, too. However, this is not the case for analog cells. The layout of most of them has to be obtained manually for specific design parameters, which are calculated during behavioral synthesis. Layout automation of analog cells is given a lot of consideration in [10].

The generation of the layout of the parameterized analog cells (like the circuits in Figures 5.5, 5.6, 5.7, 5.8, 5.12, and 5.13) is a difficult and time-consuming task [9, 43]. Reference [39] describes a cell generator that supports the design process of switched-current filters. The name of the generator is Layout. The Layout is able to apply technologies of the silicon foundry AMS (Austria Mikro Systeme), especially AMS CAE and AMS CYE technologies. The tool generates nMOS and pMOS transistors for the specified channel dimensions and parameterized multioutput current mirrors for both AMS CAE and AMS CYE technologies. Current mirrors and memory cells are used to generate SI integrators, which are necessary to realize a filter. The next section shows that for the given transfer function coefficients of a filter, the actual dimensions of transistor channels in current mirrors are calculated on the basis of parameters obtained during behavioral synthesis. For these dimensions, the elaborated tool generates the layout of transistors, whereas the topology of the current mirror is fixed.

Some special features are added to the generator in order to make the cells suitable for use in a mixed-signal environment [35]. Switching transients in the digital part of the circuit cause interference to analog circuits integrated on the same chip by means of coupling through substrate. An effective way to protect sensitive devices against coupling noise is to use guard rings and to make sure that no part of a MOS transistor is less than a specified minimum distance away from the bulk contact. This minimum distance can be a design rule imposed by the foundry or can be specified by the designer. To handle this constraint, the MOS transistor can be split into a number of parallel transistors.

The cell generator is programmed in the C^{++} language. The obtained layout of a cell is described in a CIF (Caltech intermediate format) file. The CIF format has been chosen for the VLSI geometry description because it provides a common database structure. It is readable both by many research tools and by professional industrial systems. The CIF format can also be translated to any other description of the layout, such as GDS2.

8.5 CMOS FILTER COMPILERS

It was shown in the first part of this book that OTA-C, SC, and SI techniques, apart from the digital ones, are very useful in CMOS filter implementation. However, effective use of the filter design methods requires computer tools. These tools can be introduced into the existing compilers. A system containing such tools is presented in Figure 8.5. There is a gyrator–capacitor prototype circuit in the heart of the system.

Let us assume that a third-order filter (3filter) is designed with the use of this system. Our terminology allows us to call the gyrator–capacitor prototype circuit the prototype view of the filter, 3filter[prot]. Structure numbers, which will be described in the next section, form the symbolic view of the filter, 3filter[symb]. In order to operate on these views, an object-oriented programming language is used because it can easily represent a filter cell at each level of the design process. To describe a cell at a functional level, we create a new class that defines the first- and second-order structural numbers and uses the structural number algebra. The basic tools operated on this class are symb, param, tran, and sens. Symb performs the symbolic analysis of the prototype gyrator–capacitor network on the basis of a topological method (structural numbers). The next tool, param, solves the set of nonlinear algebraic equations (7.13) in order to determine the parameters of the gyrator–capacitor circuit on the basis of the given filter specifications. The output of symb, after modification, forms the input file of the module param. The programs tran and sens calculate the transfer function amplitude and perform the sensitivity analysis of the filter. An example of the sensitivity analysis that sens performs is shown in Figure 6.19. The programs, written in the C^{++} language, perform the synthesis in an interactive system.

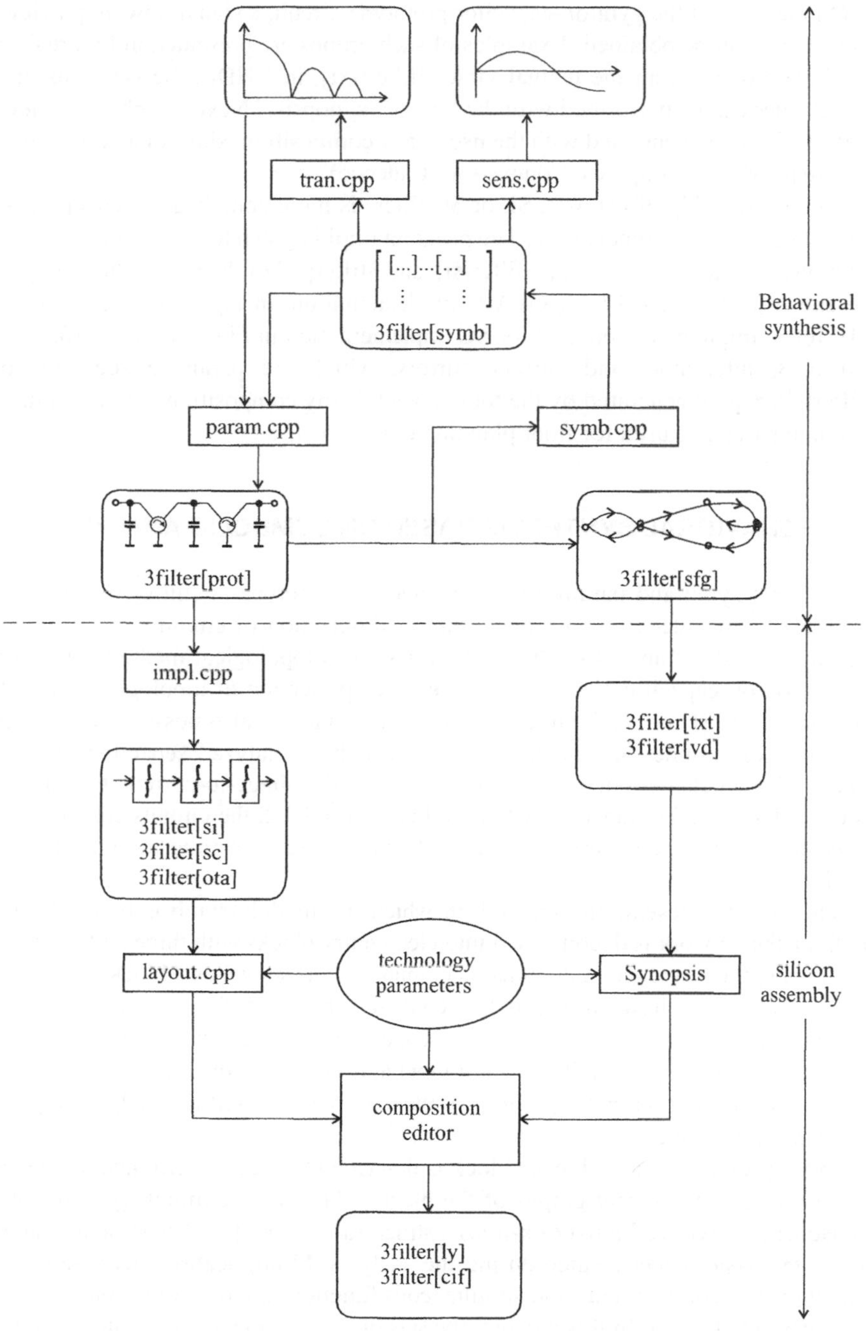

Figure 8.5 General overview of CMOS filter system.

On the basis of the gyrator–capacitor prototype circuit, a signal flow graph view, 3filter[sfg], can be obtained. Examples of such graphs are presented in Figure 6.25 and Figure 6.26. From the textual view, 3filter[txt], in VHDL, the synthesis of a digital filter can be performed with the help of "synopsis." Next, the physical view, 3filter[ly], can be generated with the use of any composition editor of a digital compiler (e.g., Mentor Graphics, Compass, or Cadence).

Silicon assembly of OTA-C, SI, or SC filters is more complicated. This process has not been fully automated yet. The program impl is a step towards automation. It generates the views 3filter[ota], 3filter[si], and 3filter[sc] of the filter. The examples are shown in Figure 6.12 for OTA-C implementation, in Figure 6.27 and Figure 7.19 for SI implementation, and for SC implementation in Figures 6.13, 6.15, 6.17, and 6.18. Integrators and current mirrors, which are parameterized cells of 3filter[si], can be generated by the tool "layout." Any composition editor of a digital compiler can be used for floor planning of these cells.

8.6 BEHAVIORAL SYNTHESIS BASED ON SYMBOLIC ANALYSIS

In the tools symb and param, the parameters of prototype circuit elements occur as symbols. In order to obtain the symbolic description of circuits, we use topological methods of analysis. This section describes topological methods of analysis of gyrator–capacitor circuits. However, the application of topological methods for analysis of SC networks in certain switching states is also possible. In this application we assume that the SC network, which is analyzed separately in each phase of a clock period, contains capacitors and voltage-to-current transducers (VCTs) described by transconductances. In Section 5.4.2, the circuits composed of capacitances and VCTs are used with reference to settling time minimization in SC filters.

This section presents the method on which the module symb is based. In this method, the network is decomposed into elementary blocks with three nodes, one of them being the reference node. Such decomposition is shown as a block graph. The elementary block, shown in Figure 8.6, contains a pair of nodes j, k, and, as optional elements, two VCTs described by transconductances g_j, g_k, two grounded capacitors C_j, C_k, a floating capacitor C_f, and input and output conductors G_j^i, G_k^i and G_j^o, G_k^o. At the assumption that $-g_j = g_k = g$, the tool symb is used for analysis of gyrator–capacitor circuits.

In the presented method, each block is described by trees determined on the basis of voltage and current graphs of the block [51]. Using terminology introduced by Bellert, we will call a list of k-trees a structural number [2]. The structural number is the basic object created during the analysis. Multiplication, derivation, and calculation of complements and simultaneous functions are operations on these objects described in [2]. In this algebra, the sets that are elements of a structural number can have more general meaning than the sets of tree branches of a graph. The elements of these sets can be structural numbers, and the objects obtained in this manner are called second category structural numbers.

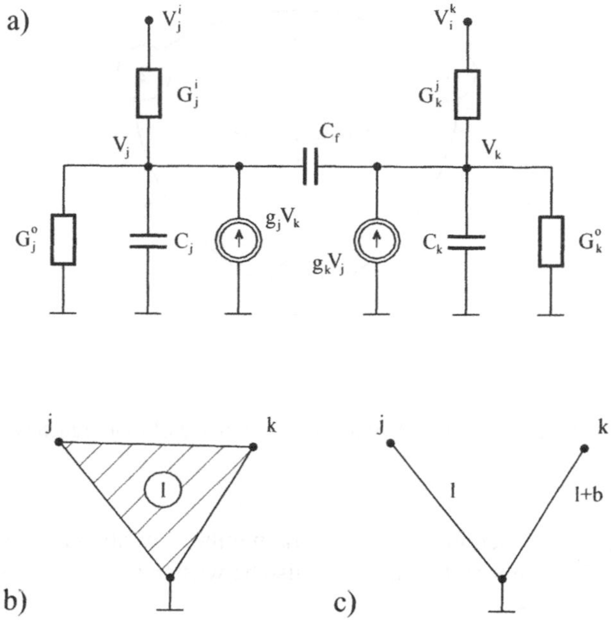

Figure 8.6 The scheme of an elementary block (a), its symbolic diagram (b), and skeleton.

The most important task in topological analysis of a circuit is the determination of trees or cotrees of a graph of this circuit. The multiplication of structural numbers is equivalent to this task. The product of two structural numbers A and B is a structural number C whose elements are obtained as sums (in the sense of the algebra of sets) of all possible combinations of disjoint elements of A and B, except for identical sets, which are removed from the number C. The occurrence of identical sets that must be removed from the list of results is called a defect of a structural number.

In order to illustrate structural number multiplication, let us consider the graph presented in Figure 8.7. Two structural numbers that denote the lists of branches incident to the nodes 1 and 2 are $A_1 = \{1\}, \{3\}, \{4\}$ and $A_2 = \{2\}, \{3\}, \{4\}$. The product of these numbers is as follows:

$$\begin{array}{cccccc} 1 & & 2 & & 1 & 2 \\ 3 & \times & 3 & = & 1 & 3 \\ 1 & & 4 & & 1 & 4 \\ & & & & 3 & 2 \\ & & & & \cancel{3} & \cancel{4} \\ & & & & 4 & 2 \\ & & & & \cancel{4} & \cancel{3} \end{array} \qquad (8.2)$$

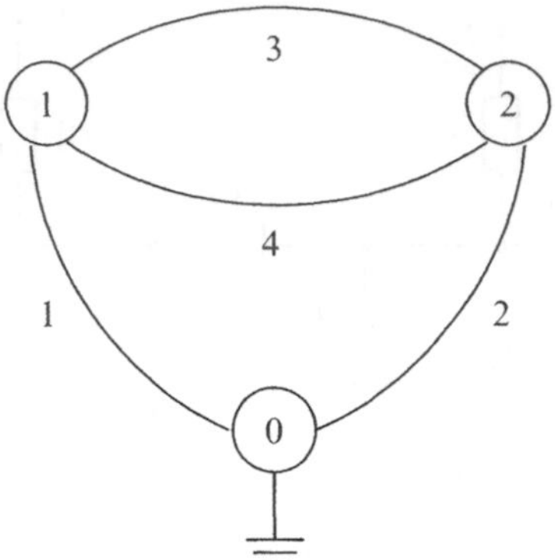

Figure 8.7 Graph of a circuit composed of four branches.

where the list of sets denoting the structural number is written in column form. The product C of the numbers A_1 and A_2 can also be written as $C = A_1 \times A_2 = \{1, 2\}, \{1, 3\}, \{1, 4\}, \{3, 2\}, \{4, 2\}$.

For the analysis of the decomposed network, the structural number of a block graph skeleton is also necessary. It is worth mentioning that for the block skeleton in Figure 8.6c, this structural number is calculated without the defect (without duplication of sets). The derivatives of this number with respect to corresponding input and output branches are calculated and for both derivatives complementary structural numbers are found. The next step of the algorithm is the calculation of the simultaneous function of the complementary numbers. The value of this function is interpreted as a second category structural number of the analyzed network. The signs in the simultaneous function (signs of k-trees) are determined on the basis of the number of sign changes in loops composed of blocks. On the basis of the second category structural number and the structural numbers of the blocks, the structural numbers corresponding to the polynomials of the numerator and the denominator of the transfer function are calculated. Let us note that in order to determine the settling time in Section 5.4.2, we have to calculate the denominator polynomial, exclusively. If the considered network is too large for symbolic analysis, then the denominator polynomial can be obtained in a semisymbolic analysis on the basis of the second category structural number.

In order to analyze a given circuit with the use of symb, it is necessary to:

1. Enumerate the nodes in the network
2. Divide the network into blocks and enumerate these blocks
3. Prepare the file with the input data in the form:

$$b \quad v$$
$$\{C_j \quad C_k \quad C_f \quad g_j \quad g_k \quad G_j^i \quad G_k^i \quad G_j^o \quad G_k^o \quad \{i\,j\,k\} \quad 0\} \qquad (8.3)$$

where

b = number of blocks,

v = number of nodes,

i, j, k = ordinal numbers of the block i and nodes j, k, respectively,

and

$$C_j|C_k|C_k|G^i|G^o = \begin{cases} 1, & C_j|\ldots|G^o \in i \\ 0, & C_j|\ldots|G^o \ni i \end{cases}$$

$$g_j|g_k = \begin{cases} 1|-1, & g \in i, \\ 0, & g \ni i \end{cases} \tag{8.4}$$

The brackets { and } in (8.3) mean that the enclosed expressions can appear many times, and the sign | in (8.4) denotes logical *or*. The values of g_j and g_k in (8.4) are equal to 1 or -1 if the direction of sources in block i is compatible with that shown in Figure 8.6a or not. For blocks composed of gyrators, the same notation is used. In (8.3) the symbols g_j and g_k are replaced by the single g, which is equal to 1 if the gyrator is directed from the node j to k and is equal to -1 if the gyrator has the opposite direction.

In order to explain the presented algorithm in detail, let us consider the gyrator–capacitor circuit of Figure 6.18a. To prepare the data describing the circuit in the form required by the program of symbolic analysis, the nodes are enumerated and the network is divided into blocks, as shown in Figure 8.8. The structure of a

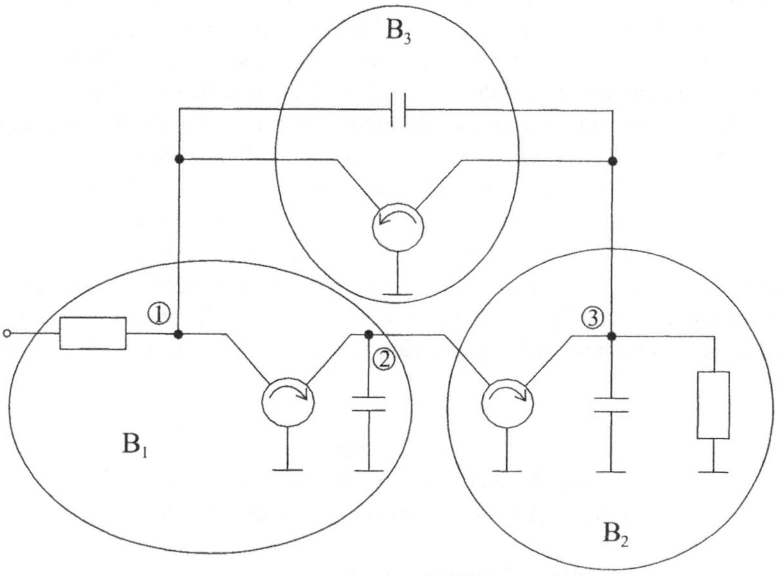

Figure 8.8 Gyrator–capacitor circuit divided into three blocks: B1, B2, and B3.

block is presented in Figure 8.6. The circuit consists of three nodes and three blocks. The first block contains nodes 1 and 2, an input conductor, a grounded capacitor, and a gyrator. The second block contains nodes 3 and 4, a conductor denoting the output node of the circuit, a capacitor, and a gyrator. The third block, between nodes 1 and 4, includes only a gyrator and a floating capacitor. Hence, using notation (8.3), the circuit can be described in the following form:

- Number of blocks $b = 3$
- Number of nodes $v = 3$
- The first block ($i = 1$), between nodes $j = 1$ and $k = 2$, is composed of C_k, G^i_j and transconductance g that corresponds to the gyrator
- The second block ($i = 2$), between nodes $j = 3$ and $k = 4$, is composed of C_k, G^o_k and g
- The third block ($i = 3$), between nodes $j = 1$, $k = 4$, contains only C_f and the gyrator directed from node k to node j

On the basis of formulae (8.3) and (8.4), the input file for symb is written as follows:

```
3   3
0   1   0   1   1   0   0   0      1   1   2   0
0   1   0   1   0   0   0   1      2   2   3   0
0   0   1  -1   0   0   0   0      3   1   3   0
```

The first row tells us that the circuit has three blocks and three nodes. The next rows describe the structure of the chosen blocks. In our example, we have different structures of all blocks. Hence, we have three rows describing the three blocks. The first three numbers (0 or 1) indicate which of the capacitors (C_j, C_k, C_f) belong to the block. The next number gives information about the presence and direction of the gyrator in the block, and the next four numbers tell us which conductors belong to the block. The next three numbers in each row indicate which blocks have this structure. In our example, in the first row we have the numbers 1 1 2 (block number 1 between nodes 1 and 2), in the second row we have 2 2 3 (block number 2 between nodes 2 and 3), and in the last row we have the numbers 3 1 3 (block number 3 between nodes 1 and 3). The last zero number denotes the end of the set of blocks with the same structure. After being run, symb reports the results in the following form:

SYMB
SYMBolic analysis of the gyrator–capacitor
multiport network based on structural numbers

r—(n + 1) * (inp*out + 1)
v—number of nodes in the network

m—number of elements
n—order of the network
inp—number of input ports
out—number of output ports
r v m
8 3 8
Description of nodes
N 1: B 1i B 3i
N 2: B 1o B 2i
N 3: B 2o B 3o
Description of blocks

	Cg1	Cg2	Cf	IG	Gi1	Gi2	Go1	Go2	Ni	No
B 1:	0	1	0	4	7	0	0	0	1	2
B 2:	0	2	0	5	0	0	0	8	2	3
B 3:	0	0	3	–6	0	0	0	0	1	3

The structural numbers of transfer function
denominator and numerators for the multiport network
(each number is divided into subsets corresponding
to coefficients of a polynomial in descending powers)

1	1	4	5
1	1	4	6
1	1	2	5
1	1	2	6
1	3	4	5
1	3	4	6
1	3	2	5
1	3	2	6

1	2	3	6
1	2	5	3
1	4	3	6
1	4	5	3
1	1	2	6
1	1	2	5
1	1	4	6
1	1	4	5

1	0	2	4
1	0	4	2
1	3	0	4
–1	–1	–1	1
–1	1	1	–1
1	2	2	3
1	2	4	0
1	4	0	3

The denominator:

1	1	2	3

1	1	2	7
1	1	3	7
1	1	3	8

1	2	4	4
1	1	7	8
1	3	4	4
–2	3	4	5
1	1	6	6
1	3	5	5

1	4	4	8
1	5	5	7

1	1	4
1	1	2
1	3	4
1	3	2

1	2	5	3	6
1	4	5	3	6
1	1	2	5	6
1	1	4	5	6

1	4	5
1	4	6
1	2	5
1	2	6

1	1	2	3	6
1	1	2	5	3
1	1	4	3	6
1	1	4	5	3

1	1	1	4
1	2	4	1

The numerator for the input branch: 7
and the output branch: 8

1	1	3	7

–1	1	6	7

1	4	5	7

In the first part of the output file, the skeleton of the block graph (Figure 8.9) and the blocks are described, where the numbers 1, . . . , 6 are associated with the inputs and outputs of the blocks:

$$1i \rightarrow 1, \quad 2i \rightarrow 2, \quad 3i \rightarrow 3, \quad 1o \rightarrow 4, \quad 2o \rightarrow 5, \quad 3o \rightarrow 6 \tag{8.5}$$

The second part contains the structural number Nv of the skeleton,

$$Nv = \begin{matrix} 1 & 4 & 5 \\ 1 & 4 & 6 \\ 1 & 2 & 5 \\ 1 & 2 & 6 \\ 3 & 4 & 5 \\ 3 & 4 & 6 \\ 3 & 2 & 5 \\ 3 & 2 & 6 \end{matrix} \tag{8.6}$$

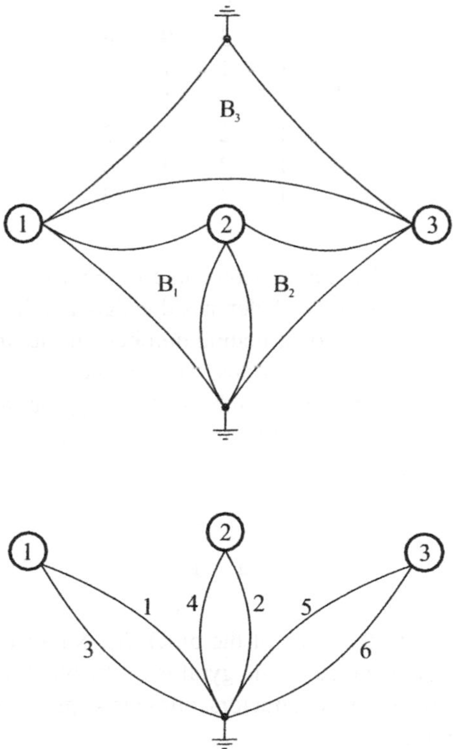

Figure 8.9 The block graph and its skeleton of the circuit from Figure 8.8.

where rows of Nv denote the trees of the skeleton. In the above array, the first column, composed of 1s, denoting the signs of trees, is omitted.

The complementary number

$$Di = \begin{matrix} 2 & 3 & 6 \\ 2 & 5 & 3 \\ 4 & 3 & 6 \\ 4 & 5 & 3 \\ 1 & 2 & 6 \\ 1 & 2 & 5 \\ 1 & 4 & 6 \\ 1 & 4 & 5 \end{matrix} \tag{8.7}$$

of Nv contains rows corresponding to cotrees of the skeleton. The second category structural number

$$S = \begin{matrix} 1 & 0 & 2 & 4 \\ 1 & 0 & 4 & 2 \\ 1 & 3 & 0 & 4 \\ -1 & -1 & -1 & 1 \\ -1 & 1 & 1 & -1 \\ 1 & 2 & 2 & 3 \\ 1 & 2 & 4 & 0 \\ 1 & 4 & 0 & 3 \end{matrix} \tag{8.8}$$

is obtained as the value of the simultaneous function Sim(Di, Di).

The first element in each row of S denotes the sign and the remaining three elements correspond to first category structural numbers of the first, second, and third blocks, respectively. In our considerations, these elements are denoted by the numbers −1, 0, 1, 2, 3, and 4, which are symbols of the appropriate sets of trees or k-trees of the corresponding block. For example, 0 in the second column of S denotes the structural number in the form:

$$\begin{matrix} 1 & 7 \\ 4 & 4 \end{matrix} \tag{8.9}$$

where rows correspond to the trees of the block B_1. Number 4 in the second row corresponds to transconductance of the gyrator in the block B_1. Since a gyrator is equivalent to two identical transconductors, number 4 appears twice. Similarly, 0 in the third column denotes

$$\begin{matrix} 5 & 5 \end{matrix} \tag{8.10}$$

corresponding to the tree of the block B_2, whereas 0 in the fourth column denotes

$$6 \quad 6 \tag{8.11}$$

corresponding to the tree of B_3. Let us note that the determinants of the admittance matrices of the blocks can be obtained on the basis of the above structural numbers as:

$$det(Y_{B1}) = sC_1G_7 + g_4^2, \qquad det(Y_{B2}) = g_5^2, \qquad det(Y_{B3}) = g_6^2 \tag{8.12}$$

The category operation on S gives the structural number of the transfer function denominator. This structural number follows the announcement "The denominator". For example, the denominator, described in the file in symbolic form, has the following algebraic form:

$$\begin{aligned} D(s) = s^3 \cdot 1.0\ C_1C_2C_3 \\ + s^2(1.0\ C_1C_2G_7 + 1.0\ C_1C_3G_7 + 1.0\ C_1C_3G_8) \\ + s^1(1.0\ C_2g_4g_4 + 1.0\ C_1G_7G_8 + 1.0\ C_3g_4g_4 \\ -2.0\ C_3g_4g_5 + 1.0\ C_1g_6g_6 + 1.0\ C_3g_5g_5) \\ + 1.0\ g_4g_4G_8 + 1.0\ g_5g_5G_7 \end{aligned} \tag{8.13}$$

where the subscripts of element symbols are taken from the structural number.

The last part of the output file shows the results of the calculation of the numerator. The structural numbers

$$Nv1 = \begin{matrix} 1 & 4 \\ 1 & 2 \\ 3 & 4 \\ 3 & 2 \end{matrix} \tag{8.14}$$

and

$$Nv2 = \begin{matrix} 4 & 5 \\ 4 & 6 \\ 2 & 5 \\ 2 & 6 \end{matrix} \tag{8.15}$$

are obtained as derivatives $Nv1 = \partial Nv/\partial 5$ and $Nv2 = \partial Nv/\partial 1$ of the structural number Nv, respectively. A derivative $\partial N/\partial n$ of the given structural number N with respect to the element n is a structural number into which all rows of N containing the element n are included after removal of this element. Rows of the derivatives $Nv1$

and *Nv*2 denote trees of the skeleton in which the branches 5 and 1 are replaced by short circuits, respectively.

The complementary numbers

$$Di1 = \begin{matrix} 2 & 5 & 3 & 6 \\ 4 & 5 & 3 & 6 \\ 1 & 4 & 5 & 6 \\ 1 & 2 & 5 & 6 \end{matrix} \tag{8.16}$$

of *Nv*1, and

$$Di2 = \begin{matrix} 1 & 2 & 3 & 6 \\ 1 & 2 & 5 & 3 \\ 1 & 4 & 3 & 6 \\ 1 & 4 & 5 & 3 \end{matrix} \tag{8.17}$$

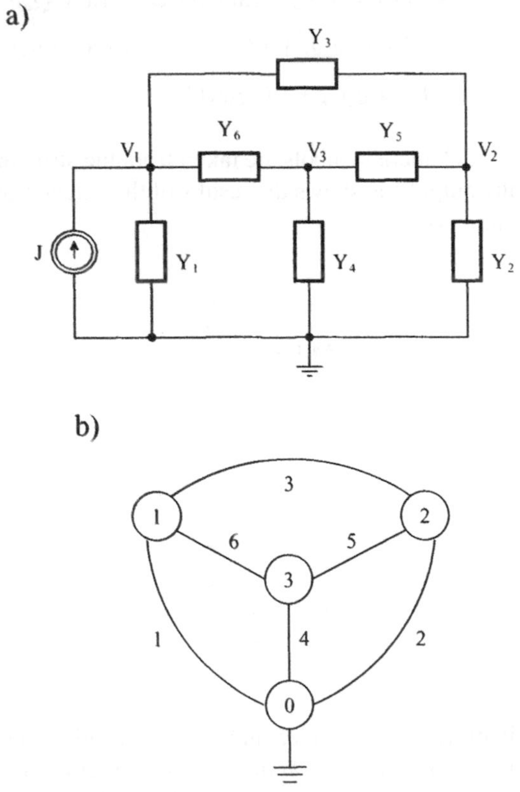

Figure 8.10 Example of a circuit (a) and its graph (b).

of $Nv2$ contain rows corresponding to cotrees of the skeleton. The second category structural number

$$Sn = \begin{matrix} 1 & 1 & 1 & 4 \\ 1 & 2 & 4 & 1 \end{matrix} \tag{8.18}$$

is obtained as the value of the simultaneous function Sim(Di1, Di2). The meaning of Sn elements is the same as in the structural number S in (8.8).

On the basis of results shown in the output file, we obtain the numerator polynomial of the transfer function in the form:

$$N(s) = s^2 \cdot 1.0\ C_1C_3G_7 - s^1 \cdot 1.0\ C_1g_6G_7 + 1.0\ g_4g_5G_7 \tag{8.19}$$

8.7 PROBLEMS

1. Calculate the structural number Nv in (8.6) as a product of the numbers $A_1 = \{1\}, \{3\}, A_2 = \{2\}, \{4\}$, and $A_3 = \{5\}, \{6\}$, denoting sets of branches incident to nodes in the graph depicted in Figure 8.9.
2. Generate all trees of the graph presented in Figure 8.10b on the basis of multiplication of structural numbers obtained as lists of branches incident to three independent nodes of the graph. Compare the result to determinant of the admittance matrix of node voltage equations of the circuit in Figure 8.10a.

References

1. N. C. Battersby and C. Toumazou, "A high-frequency fifth order switched-current bilinear elliptic lowpass filter," *IEEE J. Solid-State Circuits,* pp. 737–740, SC-29, No. 6, 1994.
2. St. Bellert, "Topological analysis and synthesis of linear systems," *J. Franklin Inst.,* 274, pp. 377–443, 1962.
3. V. Bhaskaran and K. Konstantinides, *Image and Video Compression Standards,* Kluwer Academic Publishers, Boston, MA, 1995.
4. R. W. Brodersen (editor), *Anatomy of a Silicon Compiler,* Kluwer Academic Publishers, Norwell, MA, 1992.
5. L. Burton and D. Vaughan-Pope, "Synthesis of digital ladder filters from LC filters," *IEEE Trans. Circuits Syst., CAS-23,* 6, 395–402, 1976.
6. A. P. Chandrakasan, A. Burstein, and R. W. Brodersen, "A low-power chipset for a portable multimedia I/O terminal," *IEEE Jour. of Solid-State Circuits, 29,* 12, 1415–1428, Dec. 1994.
7. A. P. Chandrakasan, S. Sheng, and R. W. Broderson, "Low-power CMOS digital design" *IEEE Journal of Solid State Circuits, 27,* 4, 473–484, 1992.
8. W. K. Chen, *Broadband Matching: Theory and Implementation,* World Scientific, Singapore, 1988.
9. J. M. Cohn, D. J. Garrod, R. A. Rutenbar, and L. R. Carley, "KOAN/ANAGRAM II: new tools for device-level analog placement and routing," *IEEE Journal of Solid State Circuits, 26,* 3, 330–342, 1991.
10. J. M. Cohn, D. J. Garrod, R. A. Rutenbar, and L. R. Carley, *Analog Device-Level Layout Automation,* Kluwer Academic Publishers, Norwell, MA, 1994.
11. F. Dufaux and F. Moscheni, "Motion estimation techniques for digital TV: A review and a new contribution," *Proc. IEEE, 83,* 858–876, 1995.

12. Badih El-Kareh, *Fundamentals of Semiconductor Processing Technology,* Kluwer Academic Publishers, Norwell, MA, 1995.

13. R. L. Fante, *Signal Analysis and Estimation: An Introduction,* Wiley, 1988.

14. A. Fettweis, "Wave digital filters: Theory and practice," *Proc. of the IEEE, 74,* 270–327, 1986.

15. I. Galton and H. T. Jensen, "Delta-sigma modulator conversion without oversampling," *IEEE Trans. Circuits Syst.—II, CAS-II–42,* 12, 773–784, 1995.

16. R. L. Geiger, P. E. Allen, and N. R. Strader, *VLSI Design Techniques for Analog and Digital Circuits,* McGraw-Hill, 1990.

17. B. Gilbert, "A precise four-quadrant multiplier with subnanosecond response," *IEEE Journal of Solid State Circuits, 3,* 4, 365–373, Dec. 1968.

18. D. Haigh and J. Everard, *GaAs Technology and Its Impact on Circuits and Systems,* Peter Peregrinus Ltd., London, 1989.

19. A. Handkiewicz, "Switched-capacitor network synthesis by extraction procedures," In *Proc. European Conference on Circuit Theory and Design, ECCTD'85,* Vol. 2, pp. 737–740, Prague 1985.

20. A. Handkiewicz, "Two-dimensional SC filter design using a gyrator-capacitor prototype," *Int. Journal of Circuit Theory and Applications, 16,* 101–105, 1988.

21. A. Handkiewicz, "Two-dimensional switched capacitor filter design system for real-time image processing," *IEEE Trans. on Circuits and Systems for Video Technology, 1,* 3, 241–246, 1991.

22. A. Handkiewicz and P. Sniatala, "Symbolic analysis approach to settling time minimization in SC networks," *Int. Journal of Circuit Theory and Applications, 23,* 357–368, 1995.

23. A. Handkiewicz, P. Sniatala, and M. Lukowiak, "High performance switched current memory cell for 2-D signal processing," In *Proc. of the European Conference of Circuit Theory and Design, ECCTD'97,* pp. 515–518, Budapest, 31. Aug.–3. Sept. 1997.

24. A. Handkiewicz, P. Sniatala, and M. Lukowiak, "Low-voltage high-performance switched current memory cell," In *Proc. Ninth Annual IEEE International ASIC Conference and Exhibit, ASIC'97,* Portland, Oregon, pp. 12–16, 7–10 Sept. 1997.

25. A. Handkiewicz, P. Sniatala, M. Lukowiak, and M. Kropidlowski, "Properties of bilinear integrators," *Electron Technology, 32,* 3, 247–250, 1999.

26. J. B. Hughes and K. W Moulding, "Switched-Current Signal Processing for Video Frequencies and Beyond," *IEEE J. of Solid-State Circuits, 28,* 1993, 314–322.

27. J. B. Hughes and K. W. Moulding, "S^2I: A switched-current technique for high performance," *Electrn. Lett., 29,* 16, 1400–1401, Aug. 5, 1993.

28. J. B. Hughes, K. W. Moulding, J. Richardson, J. Bennett, W. Redman-White, M. Bracey, and R. Singh Soin, "Automated design of switched-current filters," *IEEE Jour. of Solid-State Circuits, 31,* 7, 898–907, July 1996.

29. L. B. Jackson, *Digital Filters and Signal Processing,* Kluwer Academic Publishers, Norwell, MA, 1986.

30. G. M. Jacobs, D. J. Allstot, R. W. Brodersen, and P. R. Gray, "Design techniques for MOS switched-capacitor ladder filters," *IEEE Trans. Circuits Syst., CAS-25,* 12, 1014–1021, 1978.

31. A. K. Jain, *Fundamentals of Digital Image Processing,* Prentice Hall, 1989.

32. S. M. Kang and Y. Leblebici, *CMOS Digital Integrated Circuits: Analysis and Design*, McGraw-Hill, 1999.

33. S. Kawahito, M. Yoshida, M. Sasaki, K. Umehara, D. Miyazaki, Y. Tadokoro, K. Murata, S. Doushou, and A. Matsuzawa, "A CMOS image sensor with analog two-dimensional DCT-based compression circuits for one-chip cameras," *IEEE Journal of Solid State Circuits, 32,* 12, 2030–2041, Dec. 1997.

34. E. T. King, A. Eshraghi, I. Galton, and T. S. Fiez, "A Nyquist-rate delta-sigma A/D converter," *IEEE Journal of Solid State Circuits, 33,* 1, 45– 52, Jan. 1998.

35. K. Lampaert, G. Gielen, and W. Sansen, "A performance-driven placement tool for analog integrated circuits," *IEEE Journal of Solid State Circuits, 30,* 7, 773–780, July 1995.

36. M. S. Lee and C. Chang, "Switched-capacitor filters using the LDI and bilinear transformations," *IEEE Trans. Circuits Syst., CAS-28,* 4, 265–270, 1981.

37. M. S. Lee, G. C. Temes, C. Chang, and M. B. Ghaderi, "Bilinear switched-capacitor ladder filters," *IEEE Trans. Circuits Syst., CAS-28,* 8, 811–821, 1981.

38. M. J. Loinaz, K. J. Singh, A. J. Blanksby, D. A. Inglis, K. Azadet, and B. D. Ackland, "A 200-mW, 3. 3-V, CMOS color camera IC producing 352 × 288 24-b video at 300 frames/s," *IEEE Journal of Solid State Circuits, 33,* 12, 2092–2103, Dec. 1998.

39. M. Lukowiak, "Automated Design of Switched Current Filter Cells," (in Polish), Ph. D. dissertation, Poznan´ University of Technology, Poznan´, 2001.

40. S. K. Mendis, S. E. Kemeny, R. C. Gee, B. Pain, C. O. Staller, Q. Kim, and E. R. Fossum, "CMOS active pixel image sensor for highly integrated imaging system," *IEEE Journal of Solid State Circuits, 32,* 2, 187–197, Feb. 1997.

41. S. K. Mitra, *Analysis and Synthesis of Linear Active Networks,* Wiley, 1969.

42. S. Mutoh, T. Douseki, Y. Matsuya, T. Aoki, S. Shigematsu, and J. Yamada, "1-V power supply high-speed digital circuits technology with multithreshold-voltage CMOS," *IEEE Journal of Solid State Circuits, 30,* 8, 847–853, 1995.

43. R. Naiknaware and T. S. Fiez, "Automated hierarchical CMOS analog circuit stack generation with intramodule connectivity and matching considerations," *IEEE Journal of Solid State Circuits, 34,* 3, 304–317, 1999.

44. R. Newcomb, *Linear Multiport Synthesis,* McGraw-Hill, 1966.

45. Orchard H. J., "Inductorless filters," *Electron. Lett. 2,* 224–225, 1966.

46. Orchard H. J., Temes G. C., and Cataltepe T., "Sensitivity formulas for terminated lossless two-ports," *IEEE Trans. Circuits Syst., CAS-32,* 5, 459–466, 1985.

47. K. K. Parhi, *VLSI Digital Signal Processing Systems: Design and Implementation,* Wiley, 1999.

48. R. van de Plassche, *Integrated Analog-to-Digital and Digital-to-Analog Converters,* Kluwer Academic Publishers, Boston, MA, 1994.

49. D. B. Ribner and M. A. Copeland, "Biquad alternatives for high-frequency switched-capacitor filters," *IEEE J. Solid-State Circuits, SC-20,* 6, 1085–1094, 1985.

50. S. Sakurai and M. Ismail, *Low-Voltage CMOS Operational Amplifiers—Theory, Design and Implementation,* Kluwer Academic Publishers, Norwell, MA, 1995.

51. S. Seshu and M. B. Reed, *Linear Graphs and Electrical Networks,* Addison Wesley, 1961.

52. S. G. Smith, J. E. D. Hurwitz, M. J. Torrie, D. J. Baxter, A. A. Murray, P. Likoudis, A. J. Holmes, M. J. Panagiston, R. K. Henderson, S. Anderson, P. D. Denyer, and D. Ren-

shaw, "A single-chip CMOS 306 x 244–pixel NTSC video camera and a descendant coprocessor device," *IEEE Journal of Solid State Circuits, 33,* 12, 2104–2111, Dec. 1998.

53. C. G. Sodini, S. S. Wong, and P. K. Ko, "A framework to evaluate technology and device design enhancements for MOS integrated circuits" *IEEE Journal of Solid State Circuits, 24,* 2, 118–127, 1989.

54. N. Tan, *Switched-Current Design and Implementation of Oversampling A/D Converters,* Kluwer Academic Publishers, Boston, MA, 1997.

55. G. C. Temes and J. LaPatra, *Introduction to Circuit Synthesis and Design,* McGraw-Hill, Inc., 1977.

56. K. M. Ty and A. N. Venetsanopoulos, "A fast filter for real-time image processing," *IEEE Trans. Circuits Syst., CAS-33,* 948–957, 1986.

57. J. P. Uyemura, *Circuit Design for CMOS VLSI,* Kluwer Academic Publishers, Norwell, MA, 1992.

58. R. E. Valee and E. I. El-Masry, "A very high-frequency CMOS complementary folded cascode amplifier," *IEEE J. Solid-State Circuits, SC-29,* 2, 130–133, 1994.

59. L. Weinberg, *Network Analysis and Synthesis,* McGraw-Hill, 1962.

60. N. H. E. Weste and K. Eshraghian, *Principles of CMOS VLSI Design: A System Perspective,* Adison-Wesley, 1994.

61. H. Yoshizawa, Y. Huang, P. F. Ferguson, and G. C. Temes, "MSFET-only switched-capacitor circuits in digital CMOS technology," *IEEE J. Solid-State Circuits, SC-34,* 6, 734–747, 1999.

Index

Printed and bound by CPI Group (UK) Ltd, Croydon, CR0 4YY

27/10/2024

14580254-0001